D. C. MURDOCH
Professor of Mathematics
The University of British Columbia

Linear Algebra for All Undergraduates

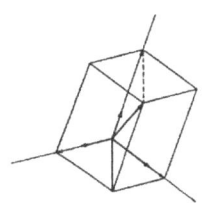

NEW YORK · JOHN WILEY & SONS, INC.
London · Chapman & Hall, Limited

Copyright © 1957
by
John Wiley & Sons, Inc.

All Rights Reserved

This book or any part thereof must not
be reproduced in any form without the
written permission of the publisher.

Preface, Recommended Reading List, New Title and All Other Revisions

@2108 Karo Maestro Blue Collar Scholar LLC (Pending)

Library of Congress Catalog Card Number: 57-8892

Printed in the United States of America

Preface

This book is intended as a text for a course in algebra and geometry suitable for third or fourth year students in mathematics and the physical sciences. It contains my answer to the frequently debated question "what should follow college algebra?" It is often claimed for books on abstract algebra that no mathematical knowledge beyond college algebra is assumed. Although this is technically correct, most students who have just completed a course in college algebra are quite unprepared for the mathematically sophisticated, abstract point of view of modern algebra. For this reason courses in abstract algebra are usually postponed to the senior year or even to the graduate school. This leaves a gap of one or two years during which many students get no further training in algebra. Others take the traditional course in theory of equations which has little application to their other studies either in mathematics or in related subjects.

It is my opinion, and also my experience, that college algebra is best followed at once by an elementary course in linear algebra. There are several reasons for this. A knowledge of linear algebra, matrices, and quadratic forms is essential to the modern student of physics and is becoming increasingly important for students in many other fields. It provides a useful background for a variety of more advanced courses, such as abstract algebra, projective geometry, tensor calculus and relativity theory, quantum mechanics, and mathematical statistics. For this reason such material should come as early as possible in the undergraduate curriculum. In addition it provides an excellent opportunity to introduce the student to a connected body of mathematical knowledge in which virtually all the proofs are given. Rigorous proofs of significant theorems are notoriously rare in elementary courses. Yet

Preface

surely mathematics *is* deductive reasoning, and surely this fact should not be concealed from the student until he is almost ready to graduate.

My object has been, therefore, to write a book that will provide for the mathematics student a smooth transition from college algebra to abstract algebra and at the same time make the basic facts of linear algebra, matrix theory, and quadratic forms available to a much larger group of students than can be expected in a full-dress course in abstract algebra. For this reason every effort has been made to keep the treatment of the various topics elementary and concrete and to introduce as few abstract ideas as possible. I have therefore defined an n-dimensional vector to be an n-tuple of numbers, a field to be a subfield of the complex numbers, and a vector space to be a vector space over the real or complex numbers. The idea of a linear transformation has been stressed and its geometric meaning emphasized, but the more abstract ideas connected with groups, fields, and rings have been virtually omitted. In spite of the importance of these I feel that they can well be left to a more advanced course for which several good books are already available. There are, of course, mathematical objections to defining a vector as an n-tuple of numbers but I am convinced that in an elementary course these are outweighed by pedagogical advantages. For students who can profit by them, abstract definitions of field and vector space are given in Appendix 1. The proofs throughout the book remain virtually unchanged if these more general definitions are adopted.

The prerequisites assumed are college algebra, including determinants, and some knowledge of three-dimensional analytic geometry. However the elementary properties of determinants have been summarized in Chapter 2 for purposes of reference, and a brief account of analytic geometry of space has been given in Appendix 2. Students who lack this background should read Appendix 2 before starting Chapter 1. The course has been successfully taught for several years by myself and my colleagues at the University of British Columbia to third year students whose mathematical background consisted of one year of calculus and one course in college algebra and analytic geometry. It has been my experience that, if the student is constantly encouraged to interpret the general theorems in ordinary three-dimensional space, the course is not found difficult even by students with the minimum prerequisite.

In the choice of subject matter I have been influenced more by what I considered useful to the student than by the desire to expound a complete mathematical theory. For this reason, as well as for reasons of space, many topics have been omitted. For example, matrices with

Preface

polynomial elements have not been considered. References have been given for the student who wishes to pursue such topics. In the arrangement of subject matter I have been concerned more for sound pedagogy than for mathematical elegance. I have been careful to provide geometric motivations for most of the algebraic theorems and have included some applications to differential equations and physical science. Problems have been provided and constitute an essential part of the course. Among these problems will be found several useful theoretical results which are not stated elsewhere in the text.

I wish to express my thanks to Professor B. N. Moyls of the University of British Columbia and to Professor M. Gweneth Humphreys of Randolph-Macon Woman's College, both of whom have read the manuscript in its preliminary stages and have offered many helpful comments; to the publishers and their consultants whose suggestions have materially improved the book; and to Mr. Bomshik Chang who assisted in checking the problems. For help with the proof reading I would like to thank my wife and my colleagues Messrs. B. Brainerd, D. W. Crowe, H. F. Davis, D. Derry, F. M. C. Goodspeed, and W. H. Simons.

D. C. MURDOCH

Vancouver, B. C.
May, 1957

Preface To The 2018 Edition

The stumbling block to becoming a mathematician appears in retrospect to have been a modest, but certainly important, topic in mathematics, commonly called linear algebra. Nothing then written about it got through to me, and little does today. I have re-read the books that puzzled me then, and I don't see how anyone can read them.

– Leonard "Jimmie" Savage, quoted in *I Want To Be A Mathematician* by Paul Halmos, page 149

This gripe by a founding father of modern Bayesian statistics is hardly an outlier on the status of linear algebra as a university course topic when he was a student (the 1930's). This probably seems unbelievable to today's students and professors of mathematics.

It's clear to anyone with even a middling knowledge of mathematics that linear algebra plays an enormous role as a unifying and simplifying principle in geometry, algebra and analysis. The central idea of analysis-both classical and modern-is the local approximation of both functions and spaces by linear transformations and/or their associated vector spaces. The classical transformations of Euclidean geometry-isometries and similarities-find their most concise and natural formulations as linear transformations of the plane. Vector and tensor analysis-without which not only multivariable calculus, but most of classical and modern physics would not exist-finds its rigorous foundations in linear algebra via vector spaces and their dual spaces. The study of linear differential equations is the study of specific function spaces and their solution subspaces. Modern differential geometry would not exist without the important vector space structures-tangent spaces, etc.-that must be erected upon abstract manifolds to generalize the local vector space structure that generates the geometry of curves and surfaces in Euclidean space. And we haven't even touched on the host of applications of linear algebra to the social and physical sciences here.

In light of the significance of the theory of linear transformations and its derived methods, a mathematics major without at least a first course in matrices, determinants, linear systems, vector spaces, linear independence, isomorphism and spectral theory-well, it seems a little like trying to make King Kong without the ape.

But it's true. Linear algebra, as most modern mathematicians think of it, simply didn't exist as a course at most universities until the 1960's.

Oh, sure, *vectors, their multivariable generalizations, tensors; matrices and their determinant functions*-were all certainly well-known and integrated into coursework at the academic level as early as the 1900's. The revolutions in physics and geometry wrought in the 19th century by the development of these basic tools were far too significant not to be taught to students. A year-long course in vector and tensor analysis was standard for students in both mathematics and physics throughout the first half of the 20th century. But the resulting calculation tools of the "arrow" and "indices" languages were tediously awkward and limited to low dimensions.

The modern coherent theory for the formalism developed quite slowly and in a piecemeal fashion. The first precise definition for a vector space seems to have originated with Giuseppe Peano in 1888. But what occupied most mathematicians and physicists at that time was the aforementioned explosive growth of vector, matrix and tensor methods in both mathematics and physics pioneered by William Gibbs, Hermann Grassman, Paul Dirac and others. The need for a rigorous foundation in terms of abstract vector spaces and linear mappings was largely driven by functional analysis and the birth of operator theory, particularly in the work of Johann Von Neumann. The unified theory as most mathematicians today would recognize it finally emerged in the 1920's with modern abstract algebra.

As slow as linear algebra's foundations were to develop, the filtering down of its' basic machinery into the mathematics curricula of universities was even slower in some regards. While linear algebra was thoroughly developed in graduate abstract algebra courses (pre-eminently van der Waerden's *Moderne Algebra)*, these courses were offered only to graduate students before World War II. While some of these universities began to offer algebra courses to advanced students in the 1930's that covered some vector space theory, the syllabi was scattershot and varied widely from program to program. These nascent courses still focused mainly on the 19th century theory of linear equations, matrices and determinants with perhaps some mention of linear mappings.

What most of us would recognize as a linear algebra textbook finally appeared in 1942 with the publication of Paul Halmos' Princeton University lecture notes delivered when he was a research assistant to Johann Von Neumann. Until Halmos wrote that original edition of *Finite Dimensional Vector Spaces,* most of the available books for these kinds of courses were not very good. The quote by Savage above perfectly encapsulates the utter incomprehensibility of those early sources and the desperation of those unfortunate students who suffered through them. Indeed, Halmos himself clearly shared his friend's exasperation in his own memory of his first exposure to linear algebra as first year graduate student taught by Henry Roy Brahana :

The algebra course was hard and I worked at it furiously....When I say furiously, I mean furiously. Brahana didn't know how to be clear, the text was Bocher's book [Introduction To Higher Algebra] (which I thought was a mess), and my dominant emotion during much of the time that I spent on the subject was exasperation reaching to anger...... I was instinctively attracted to the subject and I was upset by the needless obfuscation that kept me from getting to the heart of the matter. Vector spaces were not in the common mathematical vocabulary of Bocher's time, and linear transformations, barely mentioned, do not play an important part in his book. Orthogonal matrices are brusquely defined ($P' = P^{-1}$) in the last chapter, without any hint of their geometric significance. The classical necessary and sufficient condition for similarity (invariant factors and all that) is made to depend on the theory of "#-matrices "—they are matrices whose entries are polynomials in one variable. I organized a rump class of those who, like me, were having trouble with #-matrices ; after one of its meetings I smugly and proudly wrote in my diary, "I can teach". Some of Brahana's course was based on Dickson's [Modern Algebraic Theories], a badly written but tightly organized book, in which various canonical forms get their due share of emphasis....... somehow I survived my introduction to linear algebra. I didn't really begin to understand what the subject was about till four or five years later, after I got my Ph.D. and heard von Neumann talk about operator theory.*

What I find most interesting about this period of American mathematical education reflected in the reminisces of both Halmos and Savage was that despite the lack of good sources, clearly mathematicians at least superficially realized from the very beginning how important the subject was. If they didn't, they wouldn't have at least made the attempt to incorporate it into graduate courses. In any event, Halmos' frustration clearly translated to action once the Hungarian-born genius mentored him in the elements of linear transformations and the fog lifted. After diligently writing up von Neumann's lectures on operator algebras, Halmos vastly expanded the initial week of lecture notes, which developed the basics of linear operators on finite dimensional vector spaces, into the first version of the classic text of the same name. How his lectures on the subject came to pass has always astonished me:

The matrix business was something else again-it was, in a sense, the most exciting and most successful course I ever taught. I found von Neumann's 1939 lectures beautiful and revealing and inspiring, and in 1940 I was inspired to spread the word. The recipients of the word were to be graduate students at Princeton University. With no official pre-arrangement, I simply tacked up a card on the bulletin board in Fine Hall saying that I would offer a course called "Elementary theory of matrices ", and I proceeded to offer it. I prepared for it carefully. a goodly number of students (something like a dozen) attended regularly, and a couple of them took notes.....The notes circulated around Princeton for a while in mimeographed form. I found them invaluable; my first book, Finite-Dimensional Vector Spaces, *was based on them.*

The very first modern linear algebra course, in other words, was organized at the whim of a young postdoctoral student who volunteered to become the proverbial missionary among cannibals.

(I remember reading that as a young undergraduate and it inspiring me to work my butt off so I could get into an Ivy League graduate program-specifically so I could do what Halmos did with an "unofficial" course I wanted to teach. Boy, was I naive. Frankly, I think if *anyone* was ballsy enough to try that today at any of the

Ivies, unless he or she was either a Fields medal winner or a relative of a university donor, they'd come back to their office, find their stuff in a box outside with a pinkslip and the locks changed!)

The impact of Halmos' book on the teaching of linear algebra cannot be understated. Singlehandedly, it brought vector spaces out of graduate algebra seminars and into mainstream university coursework-albeit for advanced students. Ask any mathematician who was trained pre-1970 where they learned linear algebra and chances are, the answer will be some version of Halmos' book.

In the 1950's, a revolution began to take place in Western mathematical education, particularly America, precipitated by the beginning of the race to space between the 2 superpowers of the Cold War, the Soviet Union and America. Scientists and mathematicians, who had proven their value to society at large in their significant roles in defeating the Axis in the Second World War, achieved a heroic status with the birth of the space program. The intellectual arms race between the powers that the rapid fire achievements of NASA and their rivals in Moscow made scientists the new heroes of that generation. Children, instead of pretending to be cowboys or soldiers, pretended to be great scientists building the first rocketships to Mars. This populist canonization of the practice of mathematics and science put an unexpected and steadily increasing political pressure on both countries to heavily invest in their scientific education systems for a dramatic upgrade of the training of students at all levels. One of the immediate results of this general upgrade in university mathematical programs was the hard push to introduce linear algebra to undergraduates. At the time, Halmos was one of the very few textbooks available and programs happily seized upon it.

Unfortunately, Halmos' book emerged from a graduate seminar at one of the most advanced universities in the world. It was not a standard coursework text in the usual sense. It is a completely abstract presentation of the subject as a branch of pure algebra, with some minor applications to analysis and geometry in the exercises. Remember, when Halmos wrote his book, vector space theory was still a very new subject. It was written to collect into a single concise source from the periodical literature and treatises the essential tools graduate students needed to begin to do research. FDVS was most definitely *not* intended as an introduction to linear algebra for the average undergraduate mathematics student. This was despite the fact that when the book became popular, it was fairly clear to most in mathematical education those were the students that most needed exposure to it in order to progress to more advanced work. It was clear more accessible treatments were needed to expose students earlier if the subject was going to be effectively integrated into geometry, calculus of several variables, functional analysis and differential equations.

And that brings us, in a historically informed manner, to Murdoch's excellent text.

Murdoch's book was one of the first linear algebra textbooks that tried to give a broad presentation for well-prepared sophomores or juniors of average ability which was both mathematically careful and rich in applications. It's beautifully written, thorough and very well organized.

And the first time I read it, I was shocked it was out of print and immediately set about trying to rectify this injustice.

It emphasizes the relationship between linear algebra and the Euclidean and analytic geometry of $\#^2$ (and via isomorphism, the complex plane #) and $\#^3$ . .Murdoch completely avoids abstract vector spaces in the main text by limiting the scalar fields to # and #. I think this is a smart way to approach linear algebra in a first course. Due to the natural formulation of classical geometry in terms of vector spaces, it allows a careful presentation of the subject without recourse to abstract algebra concepts in a familiar context most students at this level should have at least a passing acquaintance with. Many later linear algebra texts took this approach-it would be interesting to see how many of the authors of those texts were aware of Murdoch's book.

Since the book was written for mathematics and science undergraduates of the late 1950's, the prerequisites are somewhat higher than for today's sophomores and juniors. Those students were expected to be familiar not only with a good standard one variable calculus course, but they were expected to have fluency in elementary algebra and standard geometry. Sadly, most of today's American freshman are far weaker at these than their predecessors. This sadly is a common problem with republished mathematics texts from that era or earlier. It's a bit like expecting most Dark Age peasants to be able to read Latin texts written at the height of the Roman Empire. Fortunately, the problem in this case is negligible because the author works quite hard to review all the needed concepts-particularly determinants and analytic geometry-to make the book self-contained.

Linear algebra-like its brother, calculus-is not well served by a completely abstract and theoretical introduction like Halmos. Its applications are as important a part of its overall structure as the underlying algebraic foundations from group and field theory are. Conversely, as in the case of calculus, a purely applied treatment which concentrates on algorithms and intuitive calculations solving problems in the sciences and omits the rigorous structure is equally misleading and distorting in the picture it provides. In linear algebra (again, as in calculus) the rigorous theory and the applications are closely intertwined and cannot really be separated without damaging understanding. Indeed, in many ways, the theory of calculus can be thought of as an application of the linear algebra of real valued functions!

The author understands this and builds the text accordingly with great care. Chapter 1 begins with vector spaces and their elementary properties, all strictly over the fields # and #. No essential structure is lost in this manner that cannot be omitted in an introduction and the generalization to abstract fields is presented briefly in an appendix. Although the book's overall emphasis is on linear transformations, Chapters 2 and 3 give a virtually complete presentation of the theory and application of linear systems of n variables, n x n matrices and determinants. In a concrete presentation, this is the natural way to go because most applications and calculations in linear algebra are done via operations on matrices i.e. linear transformations with a choice of basis. However, this actual relationship between linear transformations and matrices is not explored explicitly until later, after a great deal of experience is obtained manipulating matrices. This furnishes a great many specific examples and provides experience with calculations. Chapters 4 and 5 develop classical geometry, inner products and orthogonality relations (complements and projections) in $\#^2$ and $\#^3$. Chapter 6 finally defines and develops abstract linear transformations in $\#^2$ and $\#^3$ and the relationship to matrices under a change of basis. The critical theory of spectral analysis-i.e. the characteristic equation of a linear system and its solution producing eigenvalues as its roots and the resulting eigenvectors-is developed at length in chapter 7. This chapter is particularly noteworthy because it gives one of the clearest presentations of the deduction of the Jordan form of a matrix I've seen-a critical topic many authors struggle to give an understandable but elementary discussion of. Chapter 8 gives applications of linear algebra to the analysis of quadratic forms, an old topic that usually isn't discussed much in basic courses anymore, but is quite important in both algebra and physics. The last chapter extends the previous results to complex vector spaces, where adjoints and Hermitian matrices come into play.

Because of the sheer importance of linear algebra in today's curricula, in the last decade or so, some inexpensive quality linear algebra texts have begun to appear. Several of them will be discussed in the Further Reading section at the end of this book. That being said, the standard linear algebra texts are still quite expensive. Most of the less costly ones can't really match the more expensive standard ones in either coverage or depth.

My sincere hope is that making this book available again in Blue Collar Scholar's usual low price paperback and Ebook formats will help alleviate this problem. It matches any of the standard books in terms of coverage-indeed, it is superior to several of them in some respects. In addition, its low cost will help encourage the self-study of linear algebra by physical, social and computer science students.

My first exposure to linear algebra came many years ago in a course I basically took to push myself when I began as a double major in mathematics and biochemistry. Up until that point, I was a middling to average chemistry student who became struck with driving ambition to become an outstanding scientist, which I knew I had the ability to be. That insane semester, I very nearly gave myself a nervous breakdown by taking Introductory Linear Algebra, Differential Equations, Physical Chemistry 2, Physical Chemistry 2 Laboratory and the Chemistry Research Seminar. I was determined not to be a "loser" and let my weakness defeat me-I swore I would conquer my fears of exams. Sadly, it was during that semester that I learned how truly crippling an anxiety problem can be for a student. What I did that semester was the equivalent of locking a claustrophobe in a small dark closet-and it inevitably caught up with me. I was hospitalized the final days before finals week. My academic career never really recovered after that. As a student in both departments, I was seen as damaged goods. It was never the same after that.

The most enjoyable moments of that otherwise disastrous semester were the hours I spent learning linear algebra in Joseph Hershenov's class. One of the reasons I fell behind in my other subjects was the wonder I felt doing real mathematics for the first time. It was in that class that I first discovered the magic of proofs from pure definitions. After the first few weeks doing rote computations with matrices, row reductions into reduced row echlon form and learning how to calculate determinants, without warning, Dr. Hershenov wrote on the blackboard, "A vector space is a set with the following properties...." And he proceeded to list the 10 properties. Most of us looked at each other like he was suddenly speaking Chinese. That was quite fitting a reaction since in a sense, he was teaching us a new language-the language of mathematics. A week later, I couldn't get enough of this new language and lived in his office, where he opened more and more of this new world to me.

Professor Hershenov, after retiring due to his deteriorating health from diabetes, died alone from a violent seizure in his home. I'm sorry to say I hadn't contacted him for years when I got the sad news.

I hereby dedicate this book to this fine teacher and mathematician's memory.

Karo Maestro

New York City

April 2018

Contents

1. Vectors and Vector Spaces   1
   1. Fields   1
   2. Vectors in space and in the plane   2
   3. Vectors of order n 9
   4. Vector spaces   10
   5. Systems of linear equations   13
   6. Linear dependence of vectors   17
   7. Dimension of a vector space   20
   8. Change of basis. General coordinate systems   24
   9. Subspaces   28
   10. Inner products, length, and angle   30
   11. Application to transformations of coordinates in space   35
   12. Vectors over the complex field   37

2. Matrices, Rank, and Systems of Linear Equations   39
   13. Basic definitions   39
   14. Rank of a matrix   40
   15. Determinants   43
   16. Systems of linear equations   48

3. Further Algebra of Matrices   56
   17. Multiplication of matrices   56
   18. Inverses and zero divisors   60
   19. Addition of matrices and the laws of matrix algebra   65
   20. Some notational advantages of matrix products   66

Contents

21. Elementary transformations and determination of rank 70
22. Elementary transformations and multiplication of determinants 73
23. Multiplication of partitioned matrices 76

4. Further Geometry of Real Vector Spaces 79
24. Orthogonal complements and orthogonal projections 79
25. Equations of a subspace of $\mathscr{V}_n(\mathscr{R})$ 81
26. Volume in n-dimensional space 83

5. Transformations of Coordinates in a Vector Space 91
27. General coordinate systems 91
28. Transformation of coordinates 92
29. Normal orthogonal bases. Orthogonal matrices 94
30. Rotation and reflection of coordinate axes 96

6. Linear Transformations in a Vector Space 100
31. Definition of a linear transformation and its associated matrix relative to a given basis 100
32. Singular and nonsingular linear transformations 104
33. Examples of linear transformations 105
34. Properties of nonsingular linear transformations 109
35. The matrices of a linear transformation relative to different bases 114
36. Orthogonal transformations 118
37. Invariant subspaces and the reduction of the matrix of a linear transformation 123

7. Similar Matrices and Diagonalization Theorems 128
38. The characteristic roots and eigenvectors of a matrix 128
39. Similarity 130
40. Matrices that are similar to diagonal matrices 132
41. Systems of linear differential equations 138
42. Real symmetric matrices 144

8. Reduction of Quadratic Forms 150
43. Real quadratic forms 150
44. Quadratic forms over the complex field 153

Contents

45. Classification of quadric surfaces 156
46. Classification of quadrics by rank 160
47. Invariants 162
48. A problem in dynamics 165

9. Vector Spaces over the Complex Field 169
49. Inner products in $\mathscr{V}_n(\mathscr{C})$ 169
50. Normal orthogonal bases and unitary transformations 172
51. Hermitian matrices, forms, and transformations 174
52. Normal matrices and transformations 177
53. The spectral decomposition 180
54. The real canonical form of an orthogonal matrix 183

Appendix 1. Abstract Definition of a Vector Space 186

Appendix 2. Three Dimensional Analytic Geometry 191

References 224

Answers to Problems 225

Index 235

CHAPTER ONE

Vectors and Vector Spaces

1. FIELDS

A set of numbers is said to be *closed under addition* if the sum of any two numbers in the set is also a number of the set. The set of all rational numbers, the set of all positive integers, the set of all integral multiples of five are examples of sets that are closed under addition. Sets closed under subtraction or multiplication are defined in a similar way. A set of numbers is said to be closed under division if, whenever a and b belong to the set and $b \neq 0$, then a/b belongs to the set. A set of numbers that is closed under the four elementary operations, namely, addition, subtraction, multiplication, and division, is called a *field* of numbers. Obvious examples of fields are (*a*) the set of all complex numbers, (*b*) the set of all real numbers, (*c*) the set of all rational numbers. (The meaning of the terms rational, real, and complex as applied to numbers is discussed briefly in Appendix 2, Section A2.1).

We shall be concerned mainly with the field $\mathscr{C}$ of all complex numbers and the field $\mathscr{R}$ of all real numbers. Unless otherwise stated, small Latin letters x, y, z, a, b, c, k, etc., may be assumed to represent complex numbers. When it is necessary or desirable to restrict them to be real, this fact will be stated unless it is clear from the context. It should be clearly understood, however, that the field of complex numbers contains the field of real numbers. Thus, when we say "let a be a complex number," we do not exclude the possibility that a is real.

Exercise 1.1

1. Prove that the set of all numbers of the form $a + b\sqrt{m}$, where a and b are rational and m is a fixed integer, form a field.

2. If, in Problem 1, $\sqrt{m}$ is replaced by $\sqrt[4]{m}$, is the resulting set of numbers a field?

3. Give examples of sets of numbers that are closed under (a) addition only, (b) addition and multiplication only, (c) multiplication only, (d) addition, subtraction and multiplication, but not division.

4. Prove that any field contained in $\mathscr{C}$ that contains a number other than zero must contain all the rational numbers.

2. VECTORS IN SPACE AND IN THE PLANE

The mathematical concept of a *vector* was originally introduced to provide an adequate mathematical representation of certain physical quantities. Forces and velocities, for example, have both a magnitude and a direction associated with them and are therefore incompletely described by a single number. Logically a vector belongs with "point" and "line" among the undefined mathematical concepts that are suggested by the physical world and used to interpret it. It may be thought of as a "mathematical object" with which both a magnitude and a direction are associated. The magnitude of a vector may be any positive real number, and its direction is assumed to be defined relative to a designated coordinate system. Two vectors are said to be equal if and only if their magnitudes are equal and their directions are parallel and similarly directed. It is convenient to represent a vector by the following geometric model. Let OX_1, OX_2, OX_3 be rectangular Cartesian coordinate axes in space relative to which directions are defined in the usual way by means of direction cosines. A vector V is then uniquely determined by its magnitude r and a set of direction cosines λ, μ, ν. From the origin O draw a half-line with direction cosines λ, μ, and ν, and choose a point P on it at a distance r from the origin. The directed line segment OP is called the *geometric representation* of the vector V. It is clear that in this model two vectors have the same geometric representations if and only if they are equal; that is, their magnitudes are equal and their directions parallel and similarly directed.

The vector V is uniquely determined by the coordinates (x_1, x_2, x_3) of the terminal point P of its geometric representation. It should be noted, however, that, should the coordinate axes or the unit of measurement be changed, the three numbers (x_1, x_2, x_3) defining the vector V will in general be different. In spite of this we shall frequently write

(1) $$V = (x_1, x_2, x_3),$$

and we shall refer to the right-hand side of equation 1 as the *algebraic representation* of V relative to the chosen coordinate system. The numbers x_1, x_2, x_3 are called the coordinates of V relative to this coordinate

Vectors and Vector Spaces 3

system. Two vectors are equal if and only if their corresponding coordinates are equal. It is clear that any three real numbers (x_1, x_2, x_3) not all of which are zero define a fixed vector whose magnitude and direction can be determined from the corresponding geometric representation. It is convenient also to define a *zero vector* denoted by O whose algebraic representation is $(0, 0, 0)$. The geometric representation of O is a line segment of zero length, or simply the origin. The zero vector is therefore exceptional in that it has no direction associated with it. It has, however,

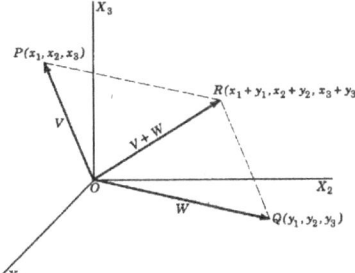

Figure 1.

magnitude 0, and hence the magnitude of a vector may be any non-negative real number. Now let

$$W = (y_1, y_2, y_3)$$

be a second vector whose geometric representation has terminal point Q. We define the sum $V + W$ to be the vector whose geometric representation is the segment OR, shown in Figure 1, which is the diagonal of the parallelogram whose sides are OP and OQ. This rule for adding vectors is known in physics as the parallelogram law. The reason for it is the experimental fact that, if the vectors V and W represent forces or velocities, then $V + W$ represents the *resultant* force or velocity. It is a simple geometric exercise (which the student should carry through) to prove that the algebraic representation of $V + W$ is given by

(2) $\qquad V + W = (x_1 + y_1, x_2 + y_2, x_3 + y_3).$

It is clear from equation 2 that addition of vectors so defined satisfies the commutative and associative laws: that is,

$$U + V = V + U$$

and $$(U + V) + W = U + (V + W).$$

The vector $V = (x_1, x_2, x_3)$ can now be written as the sum of three vectors $V_1 = (x_1, 0, 0)$, $V_2 = (0, x_2, 0)$, and $V_3 = (0, 0, x_3)$. These three

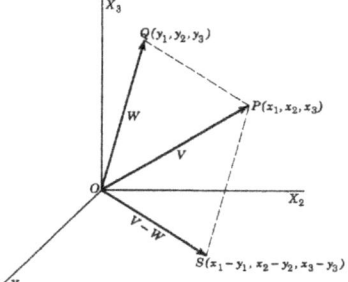

Figure 2.

vectors lie along the three coordinate axes and are called respectively the x_1-, x_2-, and x_3-components of V. The three components are uniquely determined by the vector V and the coordinate system.

The difference $V - W$ is defined to be that vector which, added to W, gives the sum V. Thus in Figure 2 the line segment OS representing $V - W$ will be determined so as to complete the parallelogram $OQPS$ of which OP is a diagonal and OQ one side. Algebraically we evidently have

(3) $$V - W = (x_1 - y_1, x_2 - y_2, x_3 - y_3).$$

Either from the geometric or the algebraic point of view the vector $V - W$ is uniquely determined when V and W are given. If we define a vector

$-W$ to be that vector, which is equal in magnitude but opposite in direction to W, we evidently have

$$-W = (-y_1, -y_2, -y_3)$$

and $W + (-W) = W - W = O$. Hence the vector $V - W$ is the same as $V + (-W)$.

If k is any non-negative real number, the vector whose magnitude is k times that of V but whose direction (if $k \neq 0$) is the same as that of V is represented by the notation kV. If k is a negative number, the notation kV will be used to mean $|k|(-V)$ or $-(|k|V)$; that is, the vector oppositely directed to V whose magnitude is $|k|$ times that of V. Hence the vector kV is defined for every real number k. It is called a *scalar multiple* of V. Here k is called a *scalar*, being a number only, in contrast to a *vector* which has both a number (its magnitude) and a direction associated with it. By use of similar triangles and the geometric representations of V and kV it is easy to show that the coordinates of kV are simply those of V multiplied by k; that is,

(4) $$kV = k(x_1, x_2, x_3) = (kx_1, kx_2, kx_3).$$

From equation 4 we find, on putting $k = 0$ and $k = -1$, that

$$(0)V = O \quad \text{and} \quad (-1)V = -V.$$

Moreover from equations 2, 3, and 4 we find that

$$k(V \pm W) = kV \pm kW.$$

We shall now summarize the rules of vector algebra. Those rules of computation which have not already been proved should be proved by the student both algebraically and geometrically.

Vector Addition

A1 The sum of two vectors V_1 and V_2 is a uniquely determined vector $V_1 + V_2$.
A2 Commutative law: $V_1 + V_2 = V_2 + V_1$.
A3 Associative law: $(V_1 + V_2) + V_3 = V_1 + (V_2 + V_3)$.
A4 There is a zero vector O having the property that $V + O = V$ for all vectors V.
A5 Every vector V has a negative $-V$ such that $V + (-V) = O$. The difference $V_1 - V_2$ may be shown to be equal to $V_1 + (-V_2)$.

Scalar Multiplication

SM1 If V is any vector and k a real number, the *scalar product* kV is a uniquely defined vector.
SM2 For every vector V, $0V = O$, $1V = V$, $(-1)V = -V$.

SM3 Associative law: $k_1(k_2V) = (k_1k_2)V$ for all scalars k_1 and k_2 and all vectors V.

SM4 Distributive laws: $k(V_1 + V_2) = kV_1 + kV_2$,
$(k_1 + k_2)V = k_1V + k_2V$,
for all scalars k, k_1, and k_2 and all vectors V_1, V_2, and V.

In addition to vector addition and scalar multiplication, it is useful to define a third operation called *inner multiplication*. Let V_1 and V_2 be any two vectors, and let $V_1 = (x_1, x_2, x_3)$ and $V_2 = (y_1, y_2, y_3)$ be their algebraic representations relative to a fixed rectangular coordinate system. The *inner product* $V_1 \cdot V_2$, also called the scalar product by some writers, is defined by the equation

(5) $\qquad V_1 \cdot V_2 = x_1y_1 + x_2y_2 + x_3y_3.$

Note that the inner product of two vectors is not a vector but a number. Since the algebraic representations of V_1 and V_2 depend on the choice of coordinate axes, it might appear that the inner product $V_1 \cdot V_2$ also depends on this choice. However, this is not true. For suppose that V_1 and V_2 have magnitudes r_1 and r_2, respectively, and suppose that the angle between their directions is θ. Considering the geometric representations of these vectors, it is known from analytic geometry (see Appendix 2, Section A2.7) that

(6) $\qquad \cos \theta = \dfrac{x_1y_1 + x_2y_2 + x_3y_3}{r_1r_2},$

and hence

$\qquad V_1 \cdot V_2 = x_1y_1 + x_2y_2 + x_3y_3 = r_1r_2 \cos \theta.$

Thus the inner product $V_1 \cdot V_2$ is the product of the magnitudes of V_1 and V_2 times the cosine of the angle between them. It therefore depends only on the vectors V_1 and V_2 and not on the particular coordinate system.†

From equation 6 we see that, if $\theta = 90°$, then $V_1 \cdot V_2 = 0$ and conversely. In this case V_1 and V_2 are said to be *orthogonal* or *perpendicular* to each other. The condition for orthogonality of two vectors, therefore, is the vanishing of their inner product. We note also that

$\qquad V_1 \cdot V_1 = x_1^2 + x_2^2 + x_3^2 = r_1^2,$

and hence the magnitude of a vector is the positive square root of its

---

† Later we shall use more general coordinate systems with oblique axes, and this statement will then require modification. However, if the inner product is defined by equation 5 in terms of the coordinates of X and Y relative to a rectangular coordinate system, then the same formula will apply for every choice of rectangular coordinate axes. If oblique axes were used the definition of $V_1 \cdot V_2$ would be changed.

SM3 Associative law: $k_1(k_2 V) = (k_1 k_2)V$ for all scalars k_1 and k_2 and all vectors V.

SM4 Distributive laws: $k(V_1 + V_2) = kV_1 + kV_2$,
$$(k_1 + k_2)V = k_1 V + k_2 V,$$
for all scalars k, k_1, and k_2 and all vectors V_1, V_2, and V.

In addition to vector addition and scalar multiplication, it is useful to define a third operation called *inner multiplication*. Let V_1 and V_2 be any two vectors, and let $V_1 = (x_1, x_2, x_3)$ and $V_2 = (y_1, y_2, y_3)$ be their algebraic representations relative to a fixed rectangular coordinate system. The *inner product* $V_1 \cdot V_2$, also called the scalar product by some writers, is defined by the equation

(5) $$V_1 \cdot V_2 = x_1 y_1 + x_2 y_2 + x_3 y_3.$$

Note that the inner product of two vectors is not a vector but a number. Since the algebraic representations of V_1 and V_2 depend on the choice of coordinate axes, it might appear that the inner product $V_1 \cdot V_2$ also depends on this choice. However, this is not true. For suppose that V_1 and V_2 have magnitudes r_1 and r_2, respectively, and suppose that the angle between their directions is θ. Considering the geometric representations of these vectors, it is known from analytic geometry (see Appendix 2, Section A2.7) that

(6) $$\cos \theta = \frac{x_1 y_1 + x_2 y_2 + x_3 y_3}{r_1 r_2},$$

and hence

$$V_1 \cdot V_2 = x_1 y_1 + x_2 y_2 + x_3 y_3 = r_1 r_2 \cos \theta.$$

Thus the inner product $V_1 \cdot V_2$ is the product of the magnitudes of V_1 and V_2 times the cosine of the angle between them. It therefore depends only on the vectors V_1 and V_2 and not on the particular coordinate system.†

From equation 6 we see that, if $\theta = 90°$, then $V_1 \cdot V_2 = 0$ and conversely. In this case V_1 and V_2 are said to be *orthogonal* or *perpendicular* to each other. The condition for orthogonality of two vectors, therefore, is the vanishing of their inner product. We note also that

$$V_1 \cdot V_1 = x_1^2 + x_2^2 + x_3^2 = r_1^2,$$

and hence the magnitude of a vector is the positive square root of its

---

† Later we shall use more general coordinate systems with oblique axes, and this statement will then require modification. However, if the inner product is defined by equation 5 in terms of the coordinates of X and Y relative to a rectangular coordinate system, then the same formula will apply for every choice of rectangular coordinate axes. If oblique axes were used the definition of $V_1 \cdot V_2$ would be changed.

inner product with itself. We shall use the notation $\|V\|$ for the magnitude of V. We can then write

$$\|V\| = \sqrt{V \cdot V}$$

and

$$\cos \theta = \frac{V_1 \cdot V_2}{\|V_1\| \, \|V_2\|}.$$

Thus both the magnitude of a vector and the angle between two vectors can be expressed in terms of inner products. Finally we note the following rules for inner multiplication which should be verified by the student.

Inner Multiplication

IM1 $V \cdot V \geq 0$ for all vectors V and $V \cdot V = 0$ implies $V = O$.
IM2 $V_1 \cdot V_2 = V_2 \cdot V_1$.
IM3 $V_1 \cdot (kV_2) = k(V_1 \cdot V_2)$.
IM4 $V_1 \cdot (V_2 + V_3) = V_1 \cdot V_2 + V_1 \cdot V_3$.

If all the vectors under consideration have directions parallel to a fixed plane, they can be represented geometrically by a two-dimensional model, that is, by directed line segments in a plane with common initial point O. Relative to a rectangular Cartesian coordinate system in this plane, with origin at O, any vector V can be represented algebraically by a pair of coordinates (x_1, x_2) which are the coordinates of the terminal point of the geometric representation of V. Addition of such "two-dimensional" vectors is again defined by the parallelogram law, and scalar multiplication and inner products are defined as in the three-dimensional case, so that

$$kV = (kx_1, kx_2) \quad \text{and} \quad V_1 \cdot V_2 = x_1 y_1 + x_2 y_2.$$

Although in the above discussion we have been careful to distinguish between a vector, its geometric representation, which is a directed line segment, and its algebraic representation, which is a symbol of the form (x_1, x_2) or (x_1, x_2, x_3), we shall, when convenient, identify a vector with either its geometric or its algebraic representation. For example, we may speak of three coplanar vectors, meaning three vectors whose geometric representations lie in one plane, or the vector (2, 1, 9) meaning the vector whose algebraic representation relative to the given coordinate system is (2, 1, 9). The tendency throughout this book will be to blend the geometric and algebraic properties of vectors so throughly that the student will not always be able to say with conviction that *this* is algebra whereas *that* is geometry.

Exercise 1.2

1. Prove equations 2 and 4 geometrically.

2. Prove the associative law $(V_1 + V_2) + V_3 = V_1 + (V_2 + V_3)$ of vector addition geometrically, both for vectors in the plane and for vectors in space.

3. If $V_1 = (x_1, x_2)$ and $V_2 = (y_1, y_2)$ are vectors in the plane, show that $V_1 \cdot V_2 = \|V_1\| \|V_2\| \cos \theta$, where θ is the angle between V_1 and V_2.

4. Find a vector V that is orthogonal to each of the vectors $(2, 1, -1)$ and $(1, 5, 2)$, and show that every scalar multiple of V is also orthogonal to these vectors.

5. Find a unit vector (i.e., a vector of magnitude 1) that is orthogonal to $(3, -4)$.

6. Find two unit vectors, orthogonal to each other, and both orthogonal to $(2, 2, 1)$.

7. If the vectors X and Y have geometric representations OP and OQ, respectively, show that the length of the projection of OQ onto the line OP is
$$\frac{|X \cdot Y|}{\|X\|}.$$

8. Find the length of the projection of the vector $(2, 7, -1)$ onto the vector $(5, 7, 4)$.

9. If V_1 and V_2 are any two vectors in space and k_1 and k_2 are any two real numbers, prove that the vector $k_1 V_1 + k_2 V_2$ lies in the same plane as V_1 and V_2.

10. If V_1 and V_2 are any two nonzero vectors in space such that V_1 is not a scalar multiple of V_2, and if V is any vector in the same plane as V_1 and V_2, prove that there exist numbers k_1 and k_2 such that $V = k_1 V_1 + k_2 V_2$.

11. If V_1 and V_2 are any two nonzero vectors in space and V_1 is not a scalar multiple of V_2, then every vector that is orthogonal to both V_1 and V_2 is orthogonal to every vector in the same plane as V_1 and V_2. Prove this algebraically, and interpret geometrically.

12. In applied mathematics and physics the *vector product* $V_1 \times V_2$ of two vectors is frequently used. It is defined as a vector whose magnitude is $\|V_1\| \|V_2\| \sin \theta$, where θ is the angle between V_1 and V_2, and whose direction is orthogonal to the plane containing V_1 and V_2 and so directed that V_1, V_2, and $V_1 \times V_2$ form a right-hand system. Prove that, if $V_1 = (x_1, x_2, x_3)$ and $V_2 = (y_1, y_2, y_3)$, then
$$V_1 \times V_2 = \pm (x_2 y_3 - x_3 y_2, x_3 y_1 - x_1 y_3, x_1 y_2 - x_2 y_1).$$
The correct sign is actually plus but this fact will be more easily proved later.

13. Prove that, if $X = (x_1, x_2, x_3)$, $Y = (y_1, y_2, y_3)$, and $Z = (z_1, z_2, z_3)$,
    (a) $X \times Y = -(Y \times X)$.
    (b) $X \times (Y + Z) = X \times Y + X \times Z$.

(c) $X \cdot (Y \times Z) = \begin{vmatrix} x_1 & x_2 & x_3 \\ y_1 & y_2 & y_3 \\ z_1 & z_2 & z_3 \end{vmatrix}$.

3. VECTORS OF ORDER n

In the last section we have tried to emphasize the fact that a vector has a dual personality; one geometric and one algebraic. Which of these is designated Dr. Jekyll and which Mr. Hyde is a matter of individual preference. However, the systematic theory that we shall develop will not be fully understood unless both the algebraic and the geometric aspects of vectors are thoroughly appreciated. Both have their advantages and their shortcomings; the algebraic aspect lends itself to easy generalization, whereas the geometric supplies concrete realizations of the abstract algebraic concepts.

From the algebraic point of view a vector in the plane is represented by a symbol of the type (x_1, x_2), and a vector in space by (x_1, x_2, x_3). There is no added difficulty in considering vectors of the form

$$(x_1, x_2, \cdots, x_n)$$

with n coordinates, and there are many good mathematical reasons for so doing. To interpret such vectors geometrically we are forced to consider spaces of n dimensions. If $n > 3$, this is outside our physical experience, but this should not worry us unduly since we are studying mathematics, not physics, and as a matter of fact the modern physicist finds many applications for the geometry of spaces of more than three dimensions. In extending our theory to vectors of this form, we shall therefore retain our geometric language since geometric interpretations of the algebraic results add to our understanding of these results even if they force us to think in terms of higher-dimensional spaces. One of the student's chief methods of study should always be to interpret our general theorems in ordinary two- and three-dimensional space, where he will usually find that they reduce to well-known theorems of plane and solid geometry.

By a *vector of order n* we shall understand a symbol of the form

$$X = (x_1, x_2, \cdots, x_n).$$

It is usual to consider only vectors in which the coordinates $x_1, x_2, \cdots, x_n$ are numbers from a fixed field $\mathscr{F}$ called the *field of scalars*. For the present the fields of scalars will be assumed to be the field $\mathscr{R}$ of all real numbers. In this case, X is called a vector over the real field or a real vector, and the term scalar is used to mean any real number. If

$$Y = (y_1, y_2, \cdots, y_n)$$

is any other real vector, we define the sum $X + Y$ to be the vector

$$X + Y = (x_1 + y_1, x_2 + y_2, \cdots, x_n + y_n).$$

The zero vector O is the vector $(0, 0, \cdots, 0)$ all of whose n coordinates are zero. Corresponding to any vector X there is a unique vector denoted by $-X$ such that $X + (-X) = O$. Evidently

$$-X = (-x_1, -x_2, \cdots, -x_n).$$

We define the difference $X - Y$ of the vectors X and Y to be the vector Z such that $Z + Y = X$. It easily follows that

$$X - Y = (x_1 - y_1, x_2 - y_2, \cdots, x_n - y_n),$$

and hence $X - Y = X + (-Y)$. If k is any scalar, we define the product kX to be the vector

$$kX = (kx_1, kx_2, \cdots, kx_n),$$

and we call this vector a scalar multiple of X. It is now easy to verify that the operations of vector addition and scalar multiplication, as we have defined them, satisfy the laws A1 through A5 and SM1 through SM4 listed in Section 2. The definition of inner products for real vectors of order n will be given later in the present chapter.

4. VECTOR SPACES

A set of vectors is said to be closed under addition if the sum of any two vectors of the set is also a vector of the set. Similarly, such a set is closed under scalar multiplication if every scalar multiple of a vector of the set is again a vector of the set.

DEFINITION. Any set of vectors over the real field that is closed under addition and scalar multiplication is called a (real) *vector space*, or *a vector space over the real field*. In any such vector space the laws A1 through A5 and SM1 through SM4 apply.

We note that, since a real vector space $\mathscr{S}$ is closed under scalar multiplication, and since $-Y = (-1)Y$, it follows that, if Y belongs to $\mathscr{S}$, $-Y$ also belongs to $\mathscr{S}$. Hence if X and Y belong to $\mathscr{S}$, $X + (-Y) = X - Y$ belongs to $\mathscr{S}$. A vector space is therefore closed under subtraction as well as under addition and scalar multiplication.

Vectors and Vector Spaces

The set of all vectors of the form

$$X = (x_1, x_2, \cdots, x_n),$$

where $x_1, x_2, \cdots, x_n$ are any real numbers, certainly constitutes a vector space since it is obviously closed under addition and scalar multiplication. Since the field of scalars here is the field $\mathscr{R}$, of all real numbers, we denote this vector space by the notation $\mathscr{V}_n(\mathscr{R})$. The space $\mathscr{V}_n(\mathscr{R})$, therefore, consists of all nth-order vectors with real coordinates. Other examples of vector spaces can easily be given. Let $X_1, X_2, \cdots, X_r$ be any r vectors of $\mathscr{V}_n(\mathscr{R})$. The vector $a_1X_1 + a_2X_2 + \cdots + a_rX_r$, where $a_1, a_2, \cdots, a_r$ are any scalars, is called a *linear combination* of the vectors $X_1, X_2, \cdots, X_r$. We can easily prove that the set of all linear combinations of a given set of vectors forms a vector space. For, if

$$V_1 = a_1X_1 + a_2X_2 + \cdots + a_rX_r$$

and

$$V_2 = b_1X_1 + b_2X_2 + \cdots + b_rX_r$$

are any two such linear combinations, then

$$V_1 + V_2 = (a_1 + b_1)X_1 + (a_2 + b_2)X_2 + \cdots + (a_r + b_r)X_r$$

and

$$kV_1 = (ka_1)X_1 + (ka_2)X_2 + \cdots + (ka_r)X_r$$

are also linear combinations of $X_1, X_2, \cdots, X_r$. The space of all linear combinations of a given set of vectors is called the space *generated* by (or *spanned* by) the given set of vectors. The space generated by a single vector X consists of all scalar multiples of X. In ordinary space or in the plane this means all vectors that lie in the same straight line as X, provided $X \neq 0$.

Consider two nonzero vectors X and Y in the space $\mathscr{V}_3(\mathscr{R})$ and the space $\mathscr{S}$ of all linear combinations $hX + kY$ for arbitrary scalars h and k. We shall assume that X is not a scalar multiple of Y, and hence, since both are nonzero, Y is not a scalar multiple of X. Since hX is a vector whose geometric representation lies in the same straight line as that of X and since kY has its geometric representation in the same straight line as that of Y, it follows from the parallelogram law that $hX + kY$ is a vector whose geometric representation lies in the same plane as X and Y. We shall show that the converse is also true; that is, every vector in the same plane as X and Y is a linear combination of X and Y. For suppose that Z is a vector in the same plane as X and Y. Through the terminal point P of Z, draw a line PA parallel to the vector Y to cut the line of the vector X in a point A. The vector with geometric representation OA is therefore a scalar multiple hX of X. Similarly, through P, draw a line PB parallel to X to cut the line of Y in B. Then the vector whose geometric representation is OB is a scalar multiple kY of Y. Moreover $Z = hX + kY$ as required. We have therefore shown that the space generated by the two vectors X and Y consists of all vectors that lie in the plane containing X and Y, the only necessary restriction on X and Y being that they do not lie in the same straight line, that is,

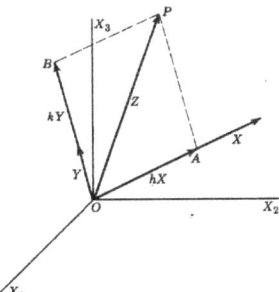

Figure 3.

one of them must not be a scalar multiple of the other. We may then state:

If X and Y are any two vectors in $\mathscr{V}_3(\mathscr{R})$ neither one of which is a scalar multiple of the other, the space generated by X and Y consists of all vectors that lie in the same plane as X and Y.

If $\mathscr{S}$ and $\mathscr{T}$ are two vector spaces over the real field and if all vectors in $\mathscr{S}$ also belong to $\mathscr{T}$, we say that $\mathscr{S}$ is a *subspace* of $\mathscr{T}$ or $\mathscr{S}$ is contained in $\mathscr{T}$, and write $\mathscr{S} \subseteq \mathscr{T}$. If, also, $\mathscr{T} \subseteq \mathscr{S}$, then $\mathscr{S}$ and $\mathscr{T}$ consist of the same vectors, and we write $\mathscr{S} = \mathscr{T}$. If $\mathscr{S} \subseteq \mathscr{T}$ but $\mathscr{S} \neq \mathscr{T}$, we say $\mathscr{S}$ is a *proper* subspace of $\mathscr{T}$, and write $\mathscr{S} \subset \mathscr{T}$. The notation $X \in \mathscr{S}$ will be used to mean that the vector X is contained in the subspace $\mathscr{S}$.

Exercise 1.3

1. If X, Y, and Z are any three vectors in space that do not lie in one plane, prove that the space generated by X, Y, and Z is the space $\mathscr{V}_3(\mathscr{R})$ of all vectors in space.

2. Prove that the zero vector O constitutes a vector space all by itself and that this "zero space," consisting of the zero vector only, is a subspace of every vector space.

3. Prove that every subspace of $\mathscr{V}_3(\mathscr{R})$ consists of either: (a) the whole space $\mathscr{V}_3(\mathscr{R})$, (b) the set of all vectors that lie in some fixed plane, (c) the set of all vectors that lie in some fixed straight line, or (d) the zero vector only.

4. Find the equation of the plane in which vectors $(2, 1, 4)$ and $(-1, 5, 2)$ lie. This may be called the equation of the space generated by these two vectors. Note that it is satisfied by the coordinates of every vector in this space.

5. Find equations of the line in which the vector $(1, 2, -6)$ lies. These may be called equations of the space generated by this vector since they are satisfied by the coordinates of every vector in this space but not by those of any vector not in the space.

6. Are the equations of a space in the sense of Problems 4 and 5 uniquely determined by the space?

7. Find a nonzero vector that belongs to the space generated by $(1, 2, -1)$ and $(3, 2, 0)$ and also to the space generated by $(2, -1, -1)$ and $(1, 0, 4)$.

8. Prove that the vector $(2, -1, 6, 4)$ belongs to the space generated by $(1, 5, -2, 1)$ and $(11, 11, 18, 19)$.

9. Show that the vectors $(2, -1, 6)$ and $(-3, 4, 1)$ generate the same space as $(-1, 3, 7,)$ and $(8, -9, 4)$.

5. SYSTEMS OF LINEAR EQUATIONS

A set of equations of the form

(7)
$$a_{11}x_1 + a_{12}x_2 + \cdots + a_{1n}x_n = c_1,$$
$$a_{21}x_1 + a_{22}x_2 + \cdots + a_{2n}x_n = c_2,$$
$$\cdots \cdots \cdots \cdots \cdots$$
$$a_{r1}x_1 + a_{r2}x_2 + \cdots + a_{rn}x_n = c_r$$

will be called a system of linear equations in the n unknowns $x_1, x_2, \cdots, x_n$. The number of equations in this system is r, and r may be less than, equal to, or greater than n. It will be assumed that the coefficients a_{ij} ($i = 1, 2, \cdots, r$ and $j = 1, 2, \cdots, n$), and the constant terms $c_1, c_2, \cdots, c_r$ are any complex numbers. They may, of course, be real but this is not assumed. Equations 7 may be written in the abbreviated form

$$\sum_{j=1}^{n} a_{ij}x_j = c_i \qquad (i = 1, 2, \cdots, r).$$

If all the constant terms $c_1, c_2, \cdots, c_r$, are zero, (7) is called a system of homogeneous linear equations; otherwise it is a system of nonhomogeneous equations. By a *solution* of this system we mean any set of values of $x_1, x_2, \cdots, x_n$ from the field of complex numbers that satisfies each of equations 7. A second system of equations

$$(8) \qquad \sum_{j=1}^{n} b_{ij}x_j = d_i \qquad (i = 1, 2, \cdots, s)$$

is said to be *equivalent* to system 7 if every solution of (7) is a solution of (8) and every solution of (8) is a solution of (7). Equivalent systems therefore have the same solutions.

A system of homogeneous linear equations in $x_1, x_2, \cdots, x_n$ always has the solution $x_1 = 0, x_2 = 0, \cdots, x_n = 0$. This is called the trivial solution. Nontrivial solutions, in which not all the unknowns are zero, may or may not exist. An important special case in which nontrivial solutions always exist is dealt with in the following theorem, which will be used frequently in what follows.

THEOREM 1.1. *A system of r linear homogeneous equations in n unknowns, with complex coefficients, always has a nontrivial solution if $r < n$. If, in addition, all the coefficients of these equations are real, they have a real nontrivial solution.*

Proof. The proof will be by mathematical induction on the number of equations. Consider the system

$$
\begin{aligned}
a_{11}x_1 + a_{12}x_2 + \cdots + a_{1n}x_n &= 0, \\
&\cdots \\
a_{r1}x_1 + a_{r2}x_2 + \cdots + a_{rn}x_n &= 0,
\end{aligned}
\qquad (9)
$$

and assume first that $r = 1$ and $n > 1$. If all the coefficients a_{1i} are zero, any choice of nonzero values of $x_1, x_2, \cdots, x_n$ will serve as a solution. If not all the coefficients are zero, by renaming the unknowns if necessary, we may assume that $a_{11} \neq 0$. By assigning arbitrary values, not all zero, to $x_2, x_3, \cdots, x_n$ and then choosing

$$x_1 = \frac{-a_{12}x_2 - a_{13}x_3 - \cdots - a_{1n}x_n}{a_{11}},$$

we get a nontrivial solution for our equation. Moreover, if all the coefficients are real and $x_2, x_3, \cdots, x_n$ are assigned real values, then x_1 will be real. The theorem is therefore proved for the case $r = 1$.

Now assume that the theorem has been proved for $r - 1$ equations in more than $r - 1$ unknowns, and consider system 9. If all the coefficients a_{ij} are zero, arbitrary values of the unknowns provide a solution. We assume therefore that at least one coefficient is not zero, and by reordering the equations and renaming the unknowns if necessary we may assume without loss of generality that $a_{11} \neq 0$. Denote the left-hand sides of equations 9 by $L_1, L_2, \cdots, L_r$, respectively. Using the first equation to eliminate x_1 from each of the remaining equations in the usual way, we find that the system

$$L_1 = 0,$$
$$L_2 - \frac{a_{21}}{a_{11}} L_1 = 0,$$

(10) $\quad\quad\quad\quad \cdots \cdots$

$$L_r - \frac{a_{r1}}{a_{11}} L_1 = 0$$

is equivalent to system 9. Since the last $r - 1$ equations of system 10 do not contain x_1, they comprise a system of $r - 1$ equations in the $n - 1$ unknowns $x_2, x_3, \cdots, x_n$. Since $r < n$, $r - 1 < n - 1$, and hence by our induction assumption the last $r - 1$ equations of system 10 have a nontrivial solution for $x_2, x_3, \cdots, x_n$ which is real if all the coefficients are real. If we put

$$x_1 = \frac{-a_{12}x_2 - a_{13}x_3 - \cdots - a_{1n}x_n}{a_{11}},$$

the first equation of system 10 is also satisfied and we have a nontrivial solution of system 10 and therefore of system 9. Moreover this solution is real if all the coefficients of system 9 are real. The induction is therefore complete, and the theorem is proved.

DEFINITION. If $x_1 = t_1, x_2 = t_2, \cdots, x_n = t_n$ is a solution of equations 7, the vector $(t_1, t_2, \cdots, t_n)$ is called a *solution vector* of system 7.

We now consider a system of homogeneous linear equations of the form (9) in which all the coefficients are real. By a solution vector of such a system we shall mean a real solution vector, since the nonreal solutions of a real system of equations are not usually of interest.

THEOREM 1.2. The (real) solution vectors of a set of homogeneous linear equations with real coefficients constitute a vector space that is a subspace of $\mathscr{V}_n(\mathscr{R})$.

Proof. Every real solution vector of equations 9 certainly belongs to $\mathscr{V}_n(\mathscr{R})$. It is only necessary to prove, therefore, (a) that the sum of any two solution vectors is a solution vector, and (b) that any scalar multiple of a solution vector is a solution vector.

Let $T = (t_1, t_2, \cdots, t_n)$ and $S = (s_1, s_2, \cdots, s_n)$ be two solution vectors of equations 9. Since

$$a_{11}(t_1 + s_1) + \cdots + a_{1n}(t_n + s_n) = (a_{11}t_1 + \cdots + a_{1n}t_n) + (a_{11}s_1 + \cdots + a_{1n}s_n) = 0,$$

the vector $T + S$ is a solution vector of the first equation of (9) and similarly of each of the others. Also, since

$$a_{11}(kt_1) + \cdots + a_{1n}(kt_n) = k(a_{11}t_1 + \cdots + a_{1n}t_n) = 0$$

for all scalars k, it follows that every scalar multiple of T is a solution vector of the first equation and similarly of each equation of the system.

The vector space consisting of all real solution vectors of a system of homogeneous linear equations with real coefficients is called the *solution space* of the system of equations. It consists of all real vectors, and only those, whose coordinates satisfy all equations of the system. Suppose (9) is the system under consideration, and let $\mathscr{S}$ be its solution space. Equations 9 are called *equations of the space* $\mathscr{S}$ since they are satisfied by the coordinates of all vectors in $\mathscr{S}$ but not by the coordinates of any real vector not in $\mathscr{S}$. The space $\mathscr{S}$ is uniquely determined by equations 9 but the equations of $\mathscr{S}$ are not uniquely determined by $\mathscr{S}$ since any system equivalent to equations 9 would define the same solution space.

Exercise 1.4

1. What is the solution space of the equation
$$x - 2y + z = 0?$$

2. What is the solution space of the equations
$$x - 2y + z = 0,$$
$$2x + y - z = 0?$$

3. Find a nontrivial solution vector of the equations
$$x + y + 2z = 0,$$
$$2x - 6y + 3z = 0,$$
and prove that every solution vector of these equations is a scalar multiple of it.

Vectors and Vector Spaces

4. Find two solution vectors of the equation
$$x + y + 3z = 0$$
such that one of them is not a scalar multiple of the other. Prove that every solution vector is a linear combination of these two.

5. Prove that the two systems of equations,

$$\begin{aligned} x + y - 3z &= 0, \\ 2x - y + z &= 0, \end{aligned} \quad \text{and} \quad \begin{aligned} 3x - 2z &= 0, \\ 7x - 2y &= 0, \end{aligned}$$

are equivalent.

6. If equations 9 have real coefficients, prove that every complex solution vector of this system has the form
$$(s_1 + it_1, s_2 + it_2, \cdots, s_n + it_n),$$
where $(s_1, s_2, \cdots, s_n)$ and $(t_1, t_2, \cdots, t_n)$ are real solution vectors of the system.

7. By successive application of the method used in the proof of Theorem 1.1, find nontrivial solutions of the following systems of equations:

(a) $2x - y + z - t = 0,$
$x + y - z + 4t = 0,$
$3x - y + 2z - t = 0.$

(b) $x + y + z + 2t - u = 0,$
$2x - y - z + t + 2u = 0,$
$x + 3y - 2z + t + u = 0.$

(c) $(2 + i)x - (4 + 3i)y - 2z = 0,$
$2x - 3iy + (1 + i)z = 0.$

(d) $x + 2y = 0,$
$(2 - i)x + 3iz = 0,$
$x - iy + (2 + i)w = 0.$

8. Prove that, if S is a solution vector of equations 7 and T is a solution vector of equation 9, then $S + T$ is a solution vector of equations 7.

9. Explain the geometric meaning of the result stated in Problem 8 when applied (a) to a single equation in two unknowns; (b) to a single equation in three unknowns.

6. LINEAR DEPENDENCE OF VECTORS

If $X_1, X_2, \cdots, X_r$ are r vectors of $\mathscr{V}_n(\mathscr{R})$ it may happen that they satisfy an equation of the form

(11) $$c_1 X_1 + c_2 X_2 + \cdots + c_r X_r = O,$$

where $c_1, c_2, \cdots, c_r$ are real numbers. Such an equation will be called a linear relation. It is obvious that, if $c_1 = c_2 = \cdots = c_r = 0$, any r vectors satisfy equation 11. Such a linear relation, in which all the coefficients are zero, will be called the trivial linear relation. Any linear relation in which not all the coefficients are zero will be called nontrivial. If two vectors X and Y satisfy a nontrivial linear relation, then $aX + bY = O$, where a and b are not both zero. If, for example, $a \neq 0$, then

$X = -(b/a)Y$, and we see that two vectors satisfy a nontrivial linear relation if and only if one of them is a scalar multiple of the other.

Example. Show that the three vectors $S = (1, 2, 0, 4)$, $T = (-1, 0, 5, 1)$, and $U = (1, 6, 10, 14)$ satisfy the linear relation

$$3S + 2T - U = O.$$

Definition. Any set of vectors that satisfy a nontrivial linear relation is said to be *linearly dependent*. A set of vectors that satisfy no linear relation except the trivial one is said to be *linearly independent*.

It should be noted that a set of vectors $X_1, X_2, \cdots, X_r$ is linearly dependent if and only if one of these vectors is a linear combination of the remaining $r - 1$. For if equation 11 is a nontrivial relation, at least one of the coefficients c_i is different from zero. There is no loss of generality in assuming $c_1 \neq 0$, in which case

$$X_1 = -\frac{c_2}{c_1} X_2 - \frac{c_3}{c_1} X_3 - \cdots - \frac{c_r}{c_1} X_r.$$

Conversely, if one of the vectors is a linear combination of the rest, this fact leads to a nontrivial linear relation of the form (11). Another way of stating this fact is

The vectors $X_1, X_2, \cdots, X_r$ are linearly dependent if and only if one of these vectors belongs to the space generated by the remaining $r - 1$.

To determine whether a given set of vectors is linearly dependent we must solve equation 11 for the unknown scalars $c_1, c_2, \cdots, c_r$ or at least determine whether a nontrivial solution exists. If the given vectors are

$$X_1 = (a_{11}, a_{21}, \cdots, a_{n1}),$$
$$X_2 = (a_{12}, a_{22}, \cdots, a_{n2}),$$
$$\cdots \cdots \cdots \cdots$$
$$X_r = (a_{1r}, a_{2r}, \cdots, a_{nr}),$$

we find on substitution that equation 11 is equivalent to the system of linear homogeneous equations

(12)
$$a_{11}c_1 + a_{12}c_2 + \cdots + a_{1r}c_r = 0,$$
$$\cdots \cdots \cdots \cdots$$
$$a_{n1}c_1 + a_{n2}c_2 + \cdots + a_{nr}c_r = 0.$$

Hence the determination of all linear relations of the form (11) for the given set of vectors amounts to the determination of all solutions of equations 12 for $c_1, c_2, \cdots, c_r$. This problem will be dealt with in full in Chapter 2, but in the meantime Theorem 1.1 gives us one important result. If $r > n$, equations 12 always have real nontrivial solutions for $c_1, c_2, \cdots, c_r$, and in this case, therefore, $X_1, X_2, \cdots, X_r$ are linearly dependent. We have therefore proved

THEOREM 1.3. *Any set of r vectors of $\mathscr{V}_n(\mathscr{R})$ is linearly dependent if $r > n$.*

The notion of linear dependence is a ubiquitous one in mathematics and is of very great importance. It is applied to functions as well as to vectors. A set of n functions $f_1(x), f_2(x), \cdots, f_n(x)$ of the real variable x, which are defined on an interval (a, b), are said to be linearly dependent in (a, b) if there exist real constants $c_1, c_2, \cdots, c_n$, not all equal to zero, such that

$$c_1 f_1(x) + c_2 f_2(x) + \cdots + c_n f_n(x) = 0$$

for all values of x in (a, b). This concept of linear dependence of functions is of basic importance in the theory of linear differential equations, a theory that bears a close analogy to the theory of linear algebraic equations developed in Chapter 2. From the abstract point of view adopted in Appendix 1 linear dependence of functions is actually a special case of linear dependence of vectors.

Exercise 1.

1. Find nontrivial linear relations satisfied by the following sets of vectors:
    (a) $(2, 1, 1), (3, -4, 6), (4, -9, 11)$.
    (b) $(2, 1), (-1, 3), (4, 2)$.
    (c) $(1, 0, 2, 4), (0, 1, 9, 2), (-5, 2, 8, -16)$.

2. Prove that any set of vectors of $\mathscr{V}_n(\mathscr{R})$ that contains the zero vector as one member of the set is linearly dependent.

3. Prove that three vectors in space are linearly dependent if and only if they lie in one plane.

4. Prove that the functions $1, x, x^2, \cdots, x^n$ are linearly independent in any interval (a, b).

5. If $f(x)$ is a polynomial of degree n, prove that $f(x)$ and its first n derivatives are linearly independent in any interval.

6. Prove that any $n + 2$ polynomials in x of degree less than or equal to n are linearly dependent in any interval.

7. If α, β, γ are any three distinct constants, prove that $\sin(x + \alpha), \sin(x + \beta), \sin(x + \gamma)$ are linearly dependent in any interval.

8. Prove that the functions $\sin x$, $\sin 2x$, $\sin 3x$, $\cdots$, $\sin nx$ are linearly independent in the interval $(0, \pi)$. *Hint:* assume

$$c_1 \sin x + c_2 \sin 2x + \cdots + c_n \sin nx = 0.$$

Multiply this equation by $\sin mx$, integrate over the interval 0 to π, and deduce that $c_m = 0$ for $m = 1, 2, \cdots, n$.

7. DIMENSION OF A VECTOR SPACE

In this section we shall give an algebraic definition of the dimension of a vector space which will coincide in the two- and three-dimensional cases with our intuitive geometric idea of dimension. Let $\mathscr{S}$ be any vector space consisting of vectors of order n; that is to say, let $\mathscr{S}$ be any subspace of $\mathscr{V}_n(\mathscr{R})$. By Theorem 1.3 we know that any set of vectors of $\mathscr{S}$ that contains more than n vectors is linearly dependent. It follows that there is a maximum number r of linearly independent vectors of $\mathscr{S}$ in the sense that (a) $\mathscr{S}$ contains at least one set of r linearly independent vectors, and (b) any set containing more than r vectors of $\mathscr{S}$ is linearly dependent. Under these circumstances we say that $\mathscr{S}$ is a space of dimension r. By Theorem 1.3, $r \leq n$. We state this definition formally as follows:

DEFINITION. The *dimension* of a vector space $\mathscr{S}$ is equal to the maximum number of linearly independent vectors containing in $\mathscr{S}$.

THEOREM 1.4. *The space $\mathscr{V}_n(\mathscr{R})$ has dimension n.*

Proof. By Theorem 1.3 any r vectors of $\mathscr{V}_n(\mathscr{R})$ where $r > n$, are linearly dependent, and therefore the dimension of $\mathscr{V}_n(\mathscr{R})$ cannot be greater than n. It is therefore sufficient to show that $\mathscr{V}_n(\mathscr{R})$ contains n linearly independent vectors. Consider the vectors

$$E_1 = (1, 0, 0, \cdots, 0),$$
$$E_2 = (0, 1, 0, \cdots, 0),$$
$$\cdots \cdots \cdots \cdots$$
$$E_n = (0, 0, 0, \cdots, 1),$$

where E_i has its ith coordinate equal to 1 and all the rest of its coordinates equal to 0. We then have

$$c_1 E_1 + c_2 E_2 + \cdots + c_n E_n = (c_1, c_2, \cdots, c_n) = O$$

only if $c_1 = c_2 = \cdots = c_n = 0$. Hence $E_1, E_2, \cdots, E_n$ are linearly independent, and the theorem follows.

Vectors and Vector Spaces

It follows from the definition that, if $\mathcal{T}$ is a subspace of $\mathcal{S}$, the dimension of $\mathcal{T}$ is less than or equal to that of $\mathcal{S}$. Hence all subspaces of $\mathcal{V}_n(\mathcal{R})$ have dimension less than or equal to n.

DEFINITION. If $X_1, X_2, \cdots, X_m$ are vectors of a space $\mathcal{S}$ such that every vector S of $\mathcal{S}$ can be written in the form

$$S = c_1 X_1 + c_2 X_2 + \cdots + c_m X_m,$$

then $X_1, X_2, \cdots, X_m$ is called a *generating system* of $\mathcal{S}$.

THEOREM 1.5. If the space $\mathcal{S}$ has a generating system consisting of m vectors $X_1, X_2, \cdots, X_m$, then any set of r vectors of $\mathcal{S}$, where $r > m$, is linearly dependent.

Proof. Let $Y_1, Y_2, \cdots, Y_r$ be any set of r vectors of $\mathcal{S}$. Since $X_1, X_2, \cdots, X_m$ is a generating system, we have

$$Y_1 = a_{11} X_1 + a_{21} X_2 + \cdots + a_{m1} X_m,$$
(13)
$$\cdots \cdots \cdots \cdots \cdots$$
$$Y_r = a_{1r} X_1 + a_{2r} X_2 + \cdots + a_{mr} X_m.$$

We wish to show that, if $r > m$, a nontrivial linear relation exists among $Y_1, Y_2, \cdots, Y_r$. From equations 13 we have

$$\begin{aligned}c_1 Y_1 + c_2 Y_2 + \cdots + c_r Y_r = &\ (a_{11}c_1 + a_{12}c_2 + \cdots + a_{1r}c_r)X_1 \\ &+ (a_{21}c_1 + a_{22}c_2 + \cdots + a_{2r}c_r)X_2 \\ &\ \cdots \cdots \cdots \cdots \cdots \\ &+ (a_{m1}c_1 + a_{m2}c_2 + \cdots + a_{mr}c_r)X_m.\end{aligned}$$

Hence we shall have

(14) $$c_1 Y_1 + c_2 Y_2 + \cdots + c_r Y_r = O,$$

provided $c_1, c_2, \cdots, c_r$ satisfy the equations

$$a_{11}c_1 + a_{12}c_2 + \cdots + a_{1r}c_r = 0,$$
$$a_{21}c_1 + a_{22}c_2 + \cdots + a_{2r}c_r = 0,$$
(15)
$$\cdots \cdots \cdots \cdots \cdots$$
$$a_{m1}c_1 + a_{m2}c_2 + \cdots + a_{mr}c_r = 0.$$

If $r > m$, equations 15 have a nontrivial solution for $c_1, \cdots, c_r$ by Theorem 1.1, and hence we always have in this case a nontrivial relation (14). The vectors $Y_1, Y_2, \cdots, Y_r$ are therefore linearly dependent as required.

THEOREM 1.6. *If $\mathscr{S}$ is a space of dimension m, every generating system of $\mathscr{S}$ contains m, but not more than m, linearly independent vectors.*

Proof. Let $X_1, X_2, \cdots, X_s$ be any generating system of $\mathscr{S}$. Let r be the greatest number of linearly independent vectors that can be chosen from $X_1, X_2, \cdots, X_s$. We may assume, then, that $X_1, X_2, \cdots, X_r$ are linearly independent but that $X_1, X_2, \cdots, X_r, X_{r+j}$ are linearly dependent for $j = 1, 2, \cdots, s - r$. Hence we have a nontrivial linear relation

$$(16) \qquad c_1 X_1 + c_2 X_2 + \cdots + c_r X_r + c_{r+j} X_{r+j} = 0,$$

in which $c_{r+j} \neq 0$, since otherwise equation 16 would be a nontrivial relation between $X_1, X_2, \cdots, X_r$, contrary to the linear independence of these vectors. Hence equation 16 defines X_{r+j} as a linear combination of $X_1, X_2, \cdots, X_r$ for $j = 1, 2, \cdots, s - r$. It follows, therefore, that $X_1, X_2, \cdots, X_r$ is also a generating system of $\mathscr{S}$.

Now by Theorem 1.5 any set of more than r vectors of $\mathscr{S}$ is linearly dependent. Since, by the definition of dimension, $\mathscr{S}$ certainly contains m linearly independent vectors, we have $m \leq r$. But since $\mathscr{S}$ contains r linearly independent vectors $X_1, X_2, \cdots, X_r$, we also have $r \leq m$. Hence $r = m$ and the theorem is proved.

DEFINITION. *A linearly independent generating system of a vector space $\mathscr{S}$ is called a* basis *of $\mathscr{S}$.*

THEOREM 1.7. *If $\mathscr{S}$ is a vector space of dimension m, every basis of $\mathscr{S}$ contains exactly m vectors and these are, of course, linearly independent. Conversely, any m linearly independent vectors of $\mathscr{S}$ constitute a basis of $\mathscr{S}$.*

Proof. Since every basis is a generating system, it contains m but not more than m linearly independent vectors by Theorem 1.6. Since the vectors of a basis are linearly independent, it therefore contains exactly m vectors in all.

Conversely let $X_1, X_2, \cdots, X_m$ be any m linearly independent vectors of $\mathscr{S}$, and let X be any other vector of $\mathscr{S}$. Since m is the dimension of $\mathscr{S}$, the $m + 1$ vectors $X_1, X_2, \cdots, X_m, X$ are linearly dependent and there exists a nontrivial relation of the form

$$c_1 X_1 + c_2 X_2 + \cdots + c_m X_m + c X = 0.$$

Moreover $c \neq 0$ since otherwise we would have a nontrivial linear relation among $X_1, X_2, \cdots, X_m$. Hence we have

$$X = -\frac{c_1}{c} X_1 - \frac{c_2}{c} X_2 - \cdots - \frac{c_m}{c} X_m.$$

Vectors and Vector Spaces

It follows that every vector X of $\mathscr{S}$ is a linear combination of $X_1, X_2, \cdots, X_m$, and hence these m vectors form a generating system. Since they are linearly independent, they also form a basis of $\mathscr{S}$.

THEOREM 1.8. If $E_1, E_2, \cdots, E_m$ is a basis of the vector space $\mathscr{S}$, then every vector X of $\mathscr{S}$ can be expressed in one and only one way in the form $X = c_1 E_1 + c_2 E_2 + \cdots + c_m E_m$.

Proof. That every vector X can be written as a linear combination of $E_1, E_2, \cdots, E_m$ follows from the definition of a basis. We need only show therefore that the coefficients $c_1, c_2, \cdots, c_m$ are uniquely determined by X. Suppose that

$$X = c_1 E_1 + c_2 E_2 + \cdots + c_m E_m = b_1 E_1 + b_2 E_2 + \cdots + b_m E_m.$$

Then $(c_1 - b_1)E_1 + (c_2 - b_2)E_2 + \cdots + (c_m - b_m)E_m = O$, and, since $E_1, E_2, \cdots, E_m$ are linearly independent, it follows that $c_1 = b_1, c_2 = b_2, \cdots, c_m = b_m$.

THEOREM 1.9. Let $\mathscr{S}$ be a vector space of dimension m, and let $X_1, X_2, \cdots, X_r, (r < m)$, be any r linearly independent vectors of $\mathscr{S}$. Then there exist $m - r$ vectors $X_{r+1}, X_{r+2}, \cdots, X_m$ of $\mathscr{S}$ which together with $X_1, X_2, \cdots, X_r$ constitute a basis of $\mathscr{S}$.

Proof. Since $r < m$, Theorem 1.6 tells us that $X_1, X_2, \cdots, X_r$ do not generate the whole space $\mathscr{S}$. Hence there exists a vector X_{r+1} of $\mathscr{S}$ that is not in the space generated by $X_1, X_2, \cdots, X_r$. The vectors $X_1, X_2, \cdots, X_r, X_{r+1}$ are therefore linearly independent, for, if

$$c_1 X_1 + c_2 X_2 + \cdots + c_r X_r + c_{r+1} X_{r+1} = O,$$

we must have $c_{r+1} = 0$ since otherwise X_{r+1} would belong to the space generated by $X_1, X_2, \cdots, X_r$. It then follows that $c_1 = c_2 = \cdots = c_r = 0$ since $X_1, X_2, \cdots, X_r$ are linearly independent. Now, if $r + 1 < m$, we can repeat this argument to obtain a vector X_{r+2} such that $X_1, X_2, \cdots, X_{r+2}$ are linearly independent and so on until we have m linearly independent vectors $X_1, X_2, \cdots, X_r, X_{r+1}, \cdots, X_m$. These constitute a basis of $\mathscr{S}$ by Theorem 1.7.

It follows from this theorem that, if $\mathscr{T}$ is a subspace of $\mathscr{S}$, any basis of $\mathscr{T}$ may be used to form part of a basis of $\mathscr{S}$.

Exercise 1.6

1. Show that each of the following sets of vectors is a vector space:
   (a) The set of all vectors in $\mathscr{V}_n(\mathscr{R}), n \geq 2$, whose first two coordinates are zero.
   (b) The set of all vectors in $\mathscr{V}_n(\mathscr{R}), n \geq 2$, whose first two coordinates are equal.

(c) The set of all vectors in $\mathscr{V}_n(\mathscr{R})$, $n \geq 3$, whose first three coordinates are equal.

(d) The set of all vectors in $\mathscr{V}_n(\mathscr{R})$ the sum of whose coordinates is zero.

2. Find the dimension and a basis for each of the vector spaces in Problem 1.

3. If X_1 and X_2 are any two vectors in $\mathscr{V}_n(\mathscr{R})$ and $Y_i = a_i X_1 + b_i X_2$ ($i = 1, 2, 3$), show that Y_1, Y_2, and Y_3 are linearly dependent.

4. Prove the following generalization of Problem 3: If each of the vectors $Y_1, Y_2, \cdots, Y_m$ is a linear combination of the r vectors $X_1, X_2, \cdots, X_r$, then, if $m > r$, the vectors $Y_1, Y_2, \cdots, Y_m$ are linearly dependent.

8. CHANGE OF BASIS. GENERAL COORDINATE SYSTEMS

It was proved in Theorem 1.7 that in a space of dimension n any n linearly independent vectors constitute a basis of the space. Consider the space $\mathscr{V}_n(\mathscr{R})$, and let

$$X = (x_1, x_2, \cdots, x_n)$$

be any vector of this space. If $F_1, F_2, \cdots, F_n$ is any basis, i.e., any set of n linearly independent vectors of $\mathscr{V}_n(\mathscr{R})$, X can be written in the form

$$X = y_1 F_1 + y_2 F_2 + \cdots + y_n F_n,$$

and by Theorem 1.8 the scalar coefficients $y_1, y_2, \cdots, y_n$ are uniquely determined by X. We call $(y_1, y_2, \cdots, y_n)$ the *coordinates of X relative to the basis* $F_1, F_2, \cdots, F_n$. Note that we also have

$$X = x_1 E_1 + x_2 E_2 + \cdots + x_n E_n,$$

where $E_1 = (1, 0, 0, \cdots, 0)$, $E_2 = (0, 1, 0, \cdots, 0), \cdots, E_n = (0, 0, \cdots, 1)$ so that $(x_1, x_2, \cdots, x_n)$ are the coordinates of X relative to the basis $E_1, E_2, \cdots, E_n$.

We see then that a vector X has a uniquely determined set of coordinates relative to every basis of the space $\mathscr{V}_n(\mathscr{R})$. We say that every basis of the space determines a coordinate system. The question of transformation from one coordinate system to another, that is, from one basis to another, will be discussed in detail in Chapter 5. Here we shall discuss different coordinate systems only for two- and three-dimensional space so that the geometric interpretation of a change of basis vectors as a transformation of coordinates may be understood in these familiar cases before it is extended to n-dimensional spaces.

Choose rectangular Cartesian coordinate axes in the plane, and let E_1 and E_2 be vectors of unit magnitude or "length" whose geometric representations lie along the x- and y-axis, respectively. Thus the algebraic representations of E_1 and E_2 are $(1, 0)$ and $(0, 1)$. Let $V = (x, y)$

Vectors and Vector Spaces

be any vector of $\mathscr{V}_2(\mathscr{R})$. Since $V = xE_1 + yE_2$, (x, y) are the coordinates of V relative to the basis E_1, E_2. Now let F_1 and F_2 be any other basis of $\mathscr{V}_2(\mathscr{R})$; that is, F_1, F_2 are any two linearly independent vectors, or any two vectors whose geometric representations do not lie in the same straight line. If $P(x, y)$ is the terminal point of V, we draw through P lines parallel to the vectors F_1 and F_2 to cut the lines of these vectors in

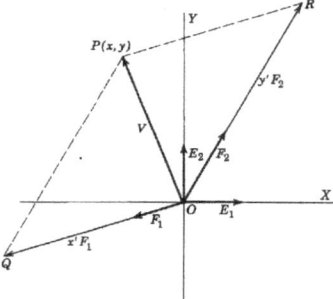

Figure 4.

Q and R (see Figure 4). The vector with geometric representation OQ is a scalar multiple of F_1, say $x'F_1$, and the vector with geometric representation OR is a scalar multiple $y'F_2$ of F_2. We then have, by the parallelogram law for addition of vectors,

$$V = x'F_1 + y'F_2,$$

and therefore (x', y') are the coordinates of V relative to the basis F_1, F_2.

If the vectors F_1 and F_2 are of unit length, the coordinates x' and y' are the lengths of OQ and OR. In this case they are often called *oblique* Cartesian coordinates of the point P relative to the coordinate axes OX' and OY' which lie along the lines OQ and OR. If F_1 and F_2 are not of unit length, (x', y') may still be regarded as oblique coordinates of P relative to the oblique axes OX' and OY', but in terms of new units of length, the length of F_1 being the new unit of the OX' axis and the length of F_2 the new unit on the OY' axis.

If the coordinates of the new basis vectors F_1 and F_2 in terms of the old basis vectors E_1 and E_2 are known, then the new coordinates (x', y') of V can be found in terms of x and y. Suppose, for example, that

(17)
$$F_1 = a_1 E_1 + a_2 E_2,$$
$$F_2 = b_1 E_1 + b_2 E_2.$$

Then we have
$$V = x'F_1 + y'F_2$$
$$+ = x'(a_1 E_1 + a_2 E_2) + y'(b_1 E_1 + b_2 E_2)$$
$$= (a_1 x' + b_1 y')E_1 + (a_2 x' + b_2 y')E_2.$$

But we also know that $V = xE_1 + yE_2$, and hence by Theorem 1.8 we get

(18)
$$x = a_1 x' + b_1 y',$$
$$y = a_2 x' + b_2 y'$$

The same method may be used to show that the coordinates of a vector relative to any two bases (i.e., any two oblique coordinate systems) are related by a pair of linear equations of the form (18).

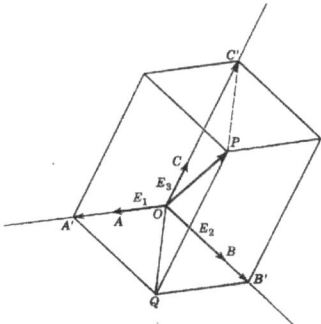

Figure 5.

In a similar way we can set up oblique Cartesian coordinates in space. Let E_1, E_2, E_3 be any three linearly independent vectors in space (i.e., any three vectors that do not lie in the same plane), represented geometrically (see Figure 5) by the line segments OA, OB, and OC. Let V

Vectors and Vector Spaces

be any vector in space, and let OP be its geometric representation. Through the terminal point P of V, pass three planes parallel to the planes of E_1 and E_2, E_2 and E_3, and E_1 and E_3, respectively. Let these three planes cut the lines of E_1, E_2, and E_3 in A', B', and C'. The vectors with geometric representations $\overrightarrow{OA'}$, $\overrightarrow{OB'}$, and $\overrightarrow{OC'}$ are then scalar multiples of E_1, E_2, and E_3 and may be written xE_1, yE_2, and zE_3. Since OP is the diagonal of a parallelepiped with edges $\overrightarrow{OA'}$, $\overrightarrow{OB'}$, and $\overrightarrow{OC'}$, we have, using $\overrightarrow{OQ}$ for the vector whose geometric representation is OQ, $\overrightarrow{OQ} = xE_1 + yE_2$ and $V = \overrightarrow{OQ} + zE_3$ and therefore

$$V = xE_1 + yE_2 + zE_3.$$

Hence (x, y, z) are the coordinates of V relative to the basis E_1, E_2, E_3 or the oblique Cartesian coordinates of the point P relative to coordinate axes which lie along the vectors E_1, E_2, and E_3. The positive directions on these coordinate axes are in the direction of the vectors E_1, E_2, and E_3, and the unit of length on each axis is the length of the corresponding basis vector.

Exercise 1.7

1. Find the coordinates of the vector $(2, 3)$ relative to the basis $F_1 = (1, -1)$ and $F_2 = (3, 5)$.

2. Find the oblique coordinates of the point $(3, 4)$ relative to the coordinate axes that have equations $y = -2x$ and $y = 2x$ in rectangular coordinates.

3. Find the coordinates of the vector $(2, 1, -6)$ relative to the basis $F_1 = (1, 1, 2)$, $F_2 = (3, -1, 0)$, $F_3 = (2, 0, -1)$.

4. If E_1, E_2, E_3 is a basis of $\mathscr{V}_3(\mathscr{R})$ and if F_1, F_2, F_3 is a second basis such that

$$F_i = a_i E_1 + b_i E_2 + c_i E_3 \qquad (i = 1, 2, 3),$$

find the equations that relate the coordinates of an arbitrary vector relative to the E-basis and the coordinates of the same vector relative to the F-basis.

5. Find the equation of the circle $x^2 + y^2 = r^2$ relative to oblique coordinate axes that make angles α and β with the original x-axis.

6. Transform the equation of the hyperbola

$$\frac{x^2}{a^2} - \frac{y^2}{b^2} = 1$$

to oblique coordinates using its asymptotes for the new coordinate axes.

9. SUBSPACES

Let $\mathscr{S}$ and $\mathscr{T}$ be any two subspaces of $\mathscr{V}_n(\mathscr{R})$, and consider the set $\mathscr{D}$ of all vectors that belong to both $\mathscr{S}$ and $\mathscr{T}$. It is easy to see that $\mathscr{D}$ must also be a vector space. For, if X belongs to $\mathscr{D}$, it belongs to both $\mathscr{S}$ and $\mathscr{T}$, and hence for every scalar k, kX belongs to both $\mathscr{S}$ and $\mathscr{T}$ and hence to $\mathscr{D}$. Similarly, if Y belongs to $\mathscr{D}$ and therefore to both $\mathscr{S}$ and $\mathscr{T}$, then $X + Y$ belongs to both $\mathscr{S}$ and $\mathscr{T}$ and therefore to $\mathscr{D}$. Hence $\mathscr{D}$ is closed under addition and scalar multiplication and is therefore a vector space.

DEFINITION. The vector space $\mathscr{D}$ which consists of all vectors common to the two spaces $\mathscr{S}$ and $\mathscr{T}$ is called the *intersection, crosscut,* or *meet* of the spaces $\mathscr{S}$ and $\mathscr{T}$. We use the notation $\mathscr{S} \cap \mathscr{T}$ for the intersection of $\mathscr{S}$ and $\mathscr{T}$.

Now consider the set $\mathscr{M}$ of all vectors of the form $S + T$ where S belongs to $\mathscr{S}$ and T belongs to $\mathscr{T}$. Since

$$(S_1 + T_1) + (S_2 + T_2) = (S_1 + S_2) + (T_1 + T_2)$$

and

$$k(S + T) = kS + kT,$$

it is clear that $\mathscr{M}$ is closed under addition and scalar multiplication, since $\mathscr{S}$ and $\mathscr{T}$ are. Hence $\mathscr{M}$ is a vector space.

DEFINITION. The space $\mathscr{M}$ consisting of all vectors $S + T$, where S belongs to $\mathscr{S}$ and T belongs to $\mathscr{T}$, is called the *sum* or the *join* of $\mathscr{S}$ and $\mathscr{T}$. We use the notation $\mathscr{S} + \mathscr{T}$ for the sum of $\mathscr{S}$ and $\mathscr{T}$.

Since the zero vector belongs to both $\mathscr{S}$ and $\mathscr{T}$, we have $S + O = S$ belongs to $\mathscr{S} + \mathscr{T}$ and $T = O + T$ belongs to $\mathscr{S} + \mathscr{T}$. Hence both $\mathscr{S}$ and $\mathscr{T}$ are subspaces of $\mathscr{S} + \mathscr{T}$. It is clear too that $\mathscr{S} \cap \mathscr{T}$ is a subspace of both $\mathscr{S}$ and $\mathscr{T}$.

THEOREM 1.10. *If $\mathscr{S}$ and $\mathscr{T}$ are subspaces of $\mathscr{V}_n(\mathscr{R})$ of dimension s and t, respectively, and if m and j are the respective dimensions of $\mathscr{S} \cap \mathscr{T}$ and $\mathscr{S} + \mathscr{T}$, then $s + t = m + j$.*

Proof. Let $X_1, X_2, \cdots, X_m$ be a basis of $\mathscr{S} \cap \mathscr{T}$. By Theorem 1.9 we can choose a basis of $\mathscr{S}$ of the form $X_1, X_2, \cdots, X_m, Y_1, \cdots, Y_{s-m}$ and a basis of $\mathscr{T}$ of the form $X_1, X_2, \cdots, X_m, Z_1, \cdots, Z_{t-m}$. It follows that every vector $S + T$ of $\mathscr{S} + \mathscr{T}$ is a linear combination of the vectors

(19) $\qquad X_1, X_2, \cdots, X_m, \quad Y_1, \cdots, Y_{s-m}, \quad Z_1, \cdots, Z_{t-m}.$

and these, therefore, form a generating system of $\mathscr{S} + \mathscr{T}$. We shall show that they are linearly independent and therefore form a basis. For let

$$a_1X_1 + \cdots + a_mX_m + b_1Y_1 + \cdots + b_{s-m}Y_{s-m} + c_1Z_1 + \cdots + c_{t-m}Z_{t-m} = 0.$$

Transposing we get

(20) $\quad a_1X_1 + \cdots + a_mX_m + \cdots + b_{s-m}Y_{s-m} = -c_1Z_1 - \cdots - c_{t-m}Z_{t-m}.$

Now the left-hand side of equation 20 is a vector of $\mathscr{S}$, and the right-hand side is a vector of $\mathscr{T}$. Their equality implies that they both belong to $\mathscr{S} \cap \mathscr{T}$, and therefore

$$-c_1Z_1 - \cdots - c_{t-m}Z_{t-m} = d_1X_1 + \cdots + d_mX_m.$$

But $X_1, \cdots, X_m, Z_1, \cdots, Z_{t-m}$ are linearly independent, being a basis of $\mathscr{T}$, and hence $c_1 = c_2 = \cdots = c_{t-m} = 0$. Therefore equation 20 becomes a linear relation in the linearly independent vectors $X_1, \cdots, X_m, Y_1, \cdots, Y_{s-m}$, and consequently

$$a_1 = a_2 = \cdots = a_m = b_1 = \cdots = b_{s-m} = 0.$$

The vectors (19) are therefore linearly independent and form a basis of $\mathscr{S} + \mathscr{T}$. Their number, namely, $s + t - m$, is therefore equal to the dimension j of $\mathscr{S} + \mathscr{T}$, and $s + t = m + j$ as required.

Exercise 1.8

1. If $\mathscr{V}_n(\mathscr{R})$ has subspaces $\mathscr{S}$ and $\mathscr{T}$, with dimensions s and t, which have no nonzero vector in common, what restriction must s and t satisfy?

2. If $\mathscr{S}$ is the space generated by the vectors $(1, 2, -1)$ and $(3, 0, 1)$ and $\mathscr{T}$ is the space generated by $(-1, 1, 0)$ and $(2, 1, 3)$, find a nonzero vector that belongs to $\mathscr{S} \cap \mathscr{T}$.

3. Prove that, if $\mathscr{S}$ is a proper subspace of $\mathscr{T}$, the dimension of $\mathscr{S}$ is less than that of $\mathscr{T}$.

4. Use Theorem 1.10 to prove that two planes that pass through the origin must either coincide or have a line in common.

5. If $\mathscr{S}$ is a subspace of a vector space $\mathscr{V}$, prove that there exists a subspace $\mathscr{T}$ of $\mathscr{V}$ such that $\mathscr{S} + \mathscr{T} = \mathscr{V}$ and $\mathscr{S} \cap \mathscr{T} = 0$. *Hint:* use Theorem 1.9.

6. If $\mathscr{S}, \mathscr{T}$, and $\mathscr{U}$ are subspaces of a vector space $\mathscr{V}$ and if $\mathscr{T} \subseteq \mathscr{S}$, prove that $\mathscr{S} \cap (\mathscr{T} + \mathscr{U}) = \mathscr{T} + (\mathscr{S} \cap \mathscr{U})$.

7. Investigate the geometric meaning of the result proved in Problem 6 when $\mathscr{S}$ and $\mathscr{U}$ are two-dimensional subspaces of $\mathscr{V}_3(\mathscr{R})$. Show by example that the hypothesis $\mathscr{T} \subseteq \mathscr{S}$ in Problem 6 is necessary for the truth of the conclusion.

10. INNER PRODUCTS, LENGTH, AND ANGLE

In Section 2 the inner product of two vectors in space or in the plane was defined, and it was pointed out that the magnitude, or length of a vector, and the angle between two vectors could both be evaluated in terms of inner products. We shall now extend our definition of inner product to vectors of $\mathscr{V}_n(\mathscr{R})$ and then use this definition to give suitable meaning to the concepts of "length of a vector" and the "angle between two vectors" in $\mathscr{V}_n(\mathscr{R})$.

DEFINITION. If $X = (x_1, x_2, \cdots, x_n)$ and $Y = (y_1, y_2, \cdots, y_n)$ are two vectors of $\mathscr{V}_n(\mathscr{R})$, the *inner product* $X \cdot Y$ of these vectors is defined by

$$X \cdot Y = x_1 y_1 + x_2 y_2 + \cdots + x_n y_n.$$

We note that $X \cdot Y$ is a scalar, not a vector, and that the operation of inner multiplication satisfies the laws IM1 through IM4 listed in Section 2.

DEFINITION. The *length* of a vector X of $\mathscr{V}_n(\mathscr{R})$ is denoted by $\|X\|$ and defined by the equation

$$\|X\| = (X \cdot X)^{1/2} = (x_1^2 + x_2^2 + \cdots + x_n^2)^{1/2}.$$

The *angle* θ between two vectors X and Y is defined by

(21) $$\cos \theta = \frac{X \cdot Y}{\|X\| \, \|Y\|}, \qquad 0 \leq \theta \leq \pi.$$

Concerning these definitions several points should be made clear. Since a vector of order n is a mathematical concept rather than a physical one, we are entitled to define its "length" in any way we please. We are guided in our choice of definition, however, by three general principles:

(*a*) Since our concept of an nth-order vector is a generalization of that of a vector in space or in the plane for which the meaning of "length" is already established, our definition should reduce in the cases $n = 2$ and $n = 3$ to the ordinary concept of length discussed in Section 2. In specializing to 2 or 3 dimensions therefore, rectangular coordinates and equal units on the coordinate axes are assumed.

(*b*) The length defined in $\mathscr{V}_n(\mathscr{R})$ should retain as far as possible the algebraic properties of length in space and in the plane.

(*c*) Our definition should lead to interesting and useful mathematical results.

Similar remarks apply to our definition of the angle between two vectors.

That our definitions of length and angle satisfy (*a*) the student can easily see by reference to Section 2. That they satisfy (*c*) he must take on trust for the present. That they satisfy (b) requires further explanation and proof. The algebraic properties of length referred to in (b) are

(1) $\|X\| \geq 0$ for every vector X and $\|X\| = 0$ if and only if $X = O$.

(2) $\|kX\| = |k| \, \|X\|$ for every real number k and every vector X.

(3) $\|X + Y\| \leq \|X\| + \|Y\|$ (the triangle inequality).

Of these, (1) expresses the essentially non-negative character of length and follows directly from the definition of $\|X\|$ as $\sqrt{x_1^2 + x_2^2 + \cdots + x_n^2}$. Similarly (2) follows from the definition and states that the length of a scalar multiple of X is equal to the length of X multiplied by the absolute value of the scalar concerned. The third property of length, when interpreted geometrically for vectors in space or in the plane, reduces (as the student should verify) to the well-known theorem that any two sides of a triangle are together greater than the third. For this reason it is called the triangle inequality. Its proof for vectors of order n will be given in Theorem 1.12.

In order that equation 21 may provide a suitable definition of the angle between the vectors X and Y it is necessary that $\dfrac{X \cdot Y}{\|X\| \, \|Y\|}$ be less than or equal to unity in absolute value, since otherwise no real angle is defined by equation 21. We shall now prove that this is always true.

THEOREM 1.11 (THE SCHWARZ INEQUALITY). If X and Y are any two vectors of $\mathscr{V}_n(\mathscr{R})$, then $|X \cdot Y| \leq \|X\| \, \|Y\|$.

Proof. The theorem is obviously true if either $X = O$ or $Y = O$. We assume therefore that X and Y are nonzero vectors. If x is any real number, we have by the rules of inner multiplication,

$$\|xX + Y\|^2 = (xX + Y) \cdot (xX + Y) = x^2\|X\|^2 + 2x(X \cdot Y) + \|Y\|^2.$$

Since $\|xX + Y\|^2$ is non-negative, we have, for all real values of x,

$$x^2\|X\|^2 + 2x(X \cdot Y) + \|Y\|^2 \geq 0.$$

But a quadratic polynomial in x is greater than or equal to zero for all real values of x if and only if its discriminant is negative or zero. Hence

$$4(X \cdot Y)^2 - 4\|X\|^2 \, \|Y\|^2 \leq 0$$

or

$$|X \cdot Y| \leq \|X\| \, \|Y\|$$

which proves the result stated. This result is known as the Schwarz inequality. It justifies the definition of $\cos \theta$ given in equation 21.

THEOREM 1.12 (THE TRIANGLE INEQUALITY). If X and Y are any two vectors of $\mathscr{V}_n(\mathscr{R})$, then $\|X + Y\| \leq \|X\| + \|Y\|$.

Proof. By the rules IM1 through IM4,

$$\begin{aligned}\|X + Y\|^2 &= (X + Y)\cdot(X + Y) = X\cdot X + 2X\cdot Y + Y\cdot Y \\ &= \|X\|^2 + 2X\cdot Y + \|Y\|^2 \\ &\leq \|X\|^2 + 2\|X\|\,\|Y\| + \|Y\|^2 \qquad \text{by Theorem 1.11} \\ &= (\|X\| + \|Y\|)^2.\end{aligned}$$

Taking positive square roots, the triangle inequality follows.

If the angle between two vectors is a right angle, they are said to be *orthogonal* or perpendicular to each other. From equations 21 we see that the condition that X and Y are orthogonal is that

$$X\cdot Y = x_1 y_1 + x_2 y_2 + \cdots + x_n y_n = 0.$$

Since $X\cdot Y = Y\cdot X$, orthogonality of vectors is a mutual relationship.

THEOREM 1.13. If a vector V is orthogonal to each of the vectors $X_1, X_2, \cdots, X_r$, it is orthogonal to every vector in the space generated by $X_1, X_2, \cdots, X_r$.

Proof. Let $X = c_1 X_1 + c_2 X_2 + \cdots + c_r X_r$ be any linear combination of $X_1, \cdots, X_r$. By IM2 and IM3,

$$\begin{aligned}V\cdot X &= V\cdot(c_1 X_1 + c_2 X_2 + \cdots + c_r X_r) \\ &= c_1 V\cdot X_1 + c_2 V\cdot X_2 + \cdots + c_r V\cdot X_r.\end{aligned}$$

Hence, if $V\cdot X_i = 0$ for $i = 1, 2, \cdots, r$, then $V\cdot X = 0$ and the theorem is proved.

DEFINITION. Two vector spaces $\mathscr{S}$ and $\mathscr{T}$ are said to be *orthogonal* to each other if every vector of $\mathscr{S}$ is orthogonal to every vector of $\mathscr{T}$.

It is clear that, if two spaces are orthogonal, every vector in either space is orthogonal to every vector in the other so that orthogonality of spaces is a mutual relationship. If $\mathscr{S}$ has a basis $S_1, S_2, \cdots, S_r$ and $\mathscr{T}$ has a basis $T_1, T_2, \cdots, T_m$, it follows from Theorem 1.13 that a necessary and sufficient condition for $\mathscr{S}$ and $\mathscr{T}$ to be orthogonal is that each basis vector S_i is orthogonal to each basis vector T_j.

The vectors $X_1, X_2, \cdots, X_r$ are said to be *mutually orthogonal* if each is orthogonal to all other vectors of the set.

THEOREM 1.14. Any set of mutually orthogonal nonzero vectors is linearly independent.

Proof. Suppose that $X_1, X_2, \cdots, X_r$ are mutually orthogonal nonzero vectors. If
$$c_1 X_1 + c_2 X_2 + \cdots + c_r X_r = O,$$
we have, on taking inner products with X_i,
$$c_1 X_1 \cdot X_i + \cdots + c_i X_i \cdot X_i + \cdots + c_r X_r \cdot X_i = 0.$$
Since $X_i \cdot X_j = 0$ for $i \neq j$ and $X_i \neq O$, we conclude that $c_i = 0$ for $i = 1, 2, \cdots, r$.

THEOREM 1.15. *If $\mathscr{S}$ is any vector space of dimension r and $\mathscr{T}$ is a subspace of $\mathscr{S}$ of dimension $m < r$, then $\mathscr{S}$ contains a nonzero vector S that is orthogonal to $\mathscr{T}$.*

Proof. Let $T_1, T_2, \cdots, T_m$ be a basis of $\mathscr{T}$. Since $\mathscr{T}$ is of smaller dimension than $\mathscr{S}$, it follows from Theorem 1.9 that $\mathscr{S}$ contains a vector S' that is not in $\mathscr{T}$, and therefore $T_1, T_2, \cdots, T_m, S'$ are linearly independent. We let

(22) $$S = a_1 T_1 + a_2 T_2 + \cdots + a_m T_m + a S'$$

and seek values of the scalars $a_1, a_2, \cdots, a_m, a$ such that S is orthogonal to each of the vectors $T_1, T_2, \cdots, T_m$. This requires that the $m + 1$ coefficients $a_1, a_2, \cdots, a_m, a$ satisfy the m homogeneous linear equations

$$a_1 T_1 \cdot T_1 + a_2 T_1 \cdot T_2 + \cdots + a_m T_1 \cdot T_m + a T_1 \cdot S' = 0,$$
$$\cdots \cdots \cdots \cdots \cdots \cdots \cdots \cdots \cdots$$
$$a_1 T_m \cdot T_1 + a_2 T_m \cdot T_2 + \cdots + a_m T_m \cdot T_m + a T_m \cdot S' = 0.$$

By Theorem 1.1 a nontrivial solution of these equations exists, and substitution in equation 22 gives a nonzero vector S since $T_1, T_2, \cdots, T_m, S'$ are linearly independent. Since S is orthogonal to each of the basis vectors $T_1, T_2, \cdots, T_m$ it is orthogonal to the space $\mathscr{T}$, and the theorem is proved.

THEOREM 1.16. *Every vector space $\mathscr{S}$ of dimension $r > 0$ contains r, but not more than r, mutually orthogonal nonzero vectors.*

Proof. Since mutually orthogonal nonzero vectors must be linearly independent, $\mathscr{S}$ cannot contain more than r of them. If $\mathscr{S}$ does not contain r mutually orthogonal vectors, let m be the maximum number of such, and let $T_1, T_2, \cdots, T_m$ be such a mutually orthogonal set. These generate a subspace $\mathscr{T}$ of $\mathscr{S}$, of dimension $m < r$. By Theorem 1.15 there exists a nonzero vector S orthogonal to each of $T_1, T_2, \cdots, T_m$ in contradiction to the supposition that m is the maximum number of mutually orthogonal vectors of $\mathscr{S}$. Hence $\mathscr{S}$ must contain r mutually orthogonal nonzero vectors as stated in the theorem.

Linear Algebra for Undergraduates

Theorem 1.16 shows that in any vector space a basis can always be chosen that consists of mutually orthogonal vectors. If $X_1, X_2, \cdots, X_r$ is such a basis for the space $\mathscr{S}$, the vectors $E_1, E_2, \cdots, E_r$, where $E_i = \dfrac{1}{\|X_i\|} X_i$, also form a basis of $\mathscr{S}$ since they are nonzero scalar multiples of $X_1, \cdots, X_r$. Moreover $\|E_i\| = 1$, for $i = 1, 2, \cdots, r$; that is, each E_i, is a *unit vector*.

DEFINITION. A basis of a vector space that consists of mutually orthogonal unit vectors is called a *normal orthogonal basis*.

Theorem 1.16 ensures that every real vector space has a normal orthogonal basis. Such a basis can actually be constructed from an arbitrary basis by an orthogonalization process which will now be described. Suppose that $X_1, X_2, \cdots, X_r$ is any basis of a space $\mathscr{S}$, and consider the vectors $Y_1, Y_2, \cdots, Y_r$ defined by

$$Y_1 = X_1,$$

$$Y_2 = X_2 - \left(\frac{X_2 \cdot X_1}{X_1 \cdot X_1}\right) X_1,$$

$$Y_3 = X_3 - \left(\frac{X_3 \cdot Y_2}{Y_2 \cdot Y_2}\right) Y_2 - \left(\frac{X_3 \cdot Y_1}{Y_1 \cdot Y_1}\right) Y_1,$$

$$\cdots \cdots \cdots \cdots \cdots \cdots \cdots \cdots \cdots \cdots$$

$$Y_r = X_r - \left(\frac{X_r \cdot Y_{r-1}}{Y_{r-1} \cdot Y_{r-1}}\right) Y_{r-1} - \cdots - \left(\frac{X_r \cdot Y_1}{Y_1 \cdot Y_1}\right) Y_1.$$

It is easy to verify successively that $Y_2 \cdot Y_1 = 0$, $Y_3 \cdot Y_1 = Y_3 \cdot Y_2 = 0$, and, in general, $Y_i \cdot Y_{i-1} = Y_i \cdot Y_{i-2} = \cdots = Y_i \cdot Y_1 = 0$. Hence the vectors Y_i are mutually orthogonal. Moreover successive substitution shows that each Y_i has the form

$$X_i - (\text{a linear combination of } X_{i-1}, \cdots, X_1),$$

which cannot be O since the X_i are linearly independent. Hence, by Theorem 1.14, $Y_1, Y_2, \cdots, Y_m$ are linearly independent and therefore constitute a basis of $\mathscr{S}$. Now, if we replace each Y_i by the corresponding unit vector $E_i = \dfrac{Y_i}{\|Y_i\|}$, we have a normal orthogonal basis $E_1, E_2, \cdots, E_r$ for $\mathscr{S}$. The construction just described is known as the *Gram-Schmidt orthogonalization process*.

THEOREM 1.17. If $E_1, E_2, \cdots, E_s$ $(1 \leq s < r)$ are mutually orthogonal unit vectors in a space $\mathscr{S}$ of dimension r, there exist unit vectors $E_{s+1}, \cdots, E_r$ in $\mathscr{S}$ such that $E_1, \cdots, E_r$ is a normal orthogonal basis of $\mathscr{S}$.

Proof. By Theorem 1.9, $\mathscr{S}$ has a basis of the form

$$E_1, \cdots, E_s, X_{s+1}, \cdots, X_r.$$

Application of the Gram-Schmidt orthogonalization process will leave $E_1, \cdots, E_s$ unchanged. Hence the final normal orthogonal basis will include $E_1, \cdots, E_s$. This result could also be proved by successive application of Theorem 1.15.

Exercise 1.9

1. Prove that the vectors $(\tfrac{1}{3}, -\tfrac{2}{3}, -\tfrac{2}{3})$, $(\tfrac{2}{3}, -\tfrac{1}{3}, \tfrac{2}{3})$, and $(\tfrac{2}{3}, \tfrac{2}{3}, -\tfrac{1}{3})$ form a normal orthogonal basis of $\mathscr{V}_3(\mathscr{R})$.

2. Find a normal orthogonal basis of $\mathscr{V}_3(\mathscr{R})$ of which $(2/\sqrt{6}, 1/\sqrt{6}, -1/\sqrt{6})$ is one vector.

3. Prove that, if two spaces $\mathscr{S}$ and $\mathscr{T}$ are orthogonal to each other, then $\mathscr{S} \cap \mathscr{T}$ consists of the zero vector only.

4. Show that the vectors $X = (1, 4, -2)$ and $Y = (2, 1, 3)$ are orthogonal, and find a third vector orthogonal to both X and Y.

5. Find two mutually orthogonal vectors each of which is orthogonal to the space generated by $(1, 0, -1, 6)$ and $(2, 4, -1, 1)$.

6. Given the basis $(2, 0, 1)$, $(3, -1, 5)$, and $(0, 4, 2)$ for $\mathscr{V}_3(\mathscr{R})$, construct a normal orthogonal basis by means of the Gram-Schmidt orthogonalization process.

7. Give a geometric interpretation of the Gram-Schmidt orthogonalization process in $\mathscr{V}_2(\mathscr{R})$ and $\mathscr{V}_3(\mathscr{R})$.

8. Write out the proof of Theorem 1.17, using the result of Theorem 1.15 instead of the Gram-Schmidt process.

9. Show that in three-dimensional space the inequality

$$\|X + Y\| \leq \|X\| + \|Y\|$$

follows from the fact that "any two sides of a triangle are together greater than the third side."

10. Prove that, if $\|X + Y\| = \|X\| + \|Y\|$, then X and Y are linearly dependent. *Hint:* examine the proofs of Theorems 1.12 and 1.11, and see what this equality implies.

11. APPLICATION TO TRANSFORMATIONS OF COORDINATES IN SPACE

Consider a system of rectangular Cartesian coordinates in space with coordinate axes OX, OY, and OZ, and a second set of coordinate axes OX', OY', and OZ' whose direction cosines relative to OX, OY, and OZ

are, respectively, $(\lambda_1, \mu_1, \nu_1)$, $(\lambda_2, \mu_2, \nu_2)$, and $(\lambda_3, \mu_3, \nu_3)$. Let E_1, E_2, and E_3 be unit vectors along OX, OY, and OZ, and F_1, F_2, F_3 be unit vectors along OX', OY', and OZ'. Since F_1 is of unit length, its coordinates relative to E_1, E_2, E_3 are $(\lambda_1, \mu_1, \nu_1)$. Similarly F_2 and F_3 have coordinates $(\lambda_2, \mu_2, \nu_2)$ and $(\lambda_3, \mu_3, \nu_3)$ relative to the E-basis. Hence we may write

(23)
$$F_1 = \lambda_1 E_1 + \mu_1 E_2 + \nu_1 E_3,$$
$$F_2 = \lambda_2 E_1 + \mu_2 E_2 + \nu_2 E_3,$$
$$F_3 = \lambda_3 E_1 + \mu_3 E_2 + \nu_3 E_3.$$

Now let (x, y, z) be the rectangular coordinates relative to OX, OY, OZ of any point P, and let (x', y', z') be the coordinates of P relative to OX', OY', OZ'. Let V be the vector whose terminal point is P. We then have by equations 23

$$V = xE_1 + yE_2 + zE_3 = x'F_1 + y'F_2 + z'F_3$$
$$= (\lambda_1 x' + \lambda_2 y' + \lambda_3 z')E_1 + (\mu_1 x' + \mu_2 y' + \mu_3 z')E_2 + (\nu_1 x' + \nu_2 y' + \nu_3 z')E_3.$$

By Theorem 1.8, therefore,

(24)
$$x = \lambda_1 x' + \lambda_2 y' + \lambda_3 z',$$
$$y = \mu_1 x' + \mu_2 y' + \mu_3 z',$$
$$z = \nu_1 x' + \nu_2 y' + \nu_3 z'.$$

Equations 24 are the equations of the transformation of coordinates from the x-, y-, z-axes to the x'-, y'-, z'-axes.

An important special case occurs when the basis vectors E_1, E_2, E_3 can be brought into coincidence with F_1, F_2, F_3 by means of a rigid rotation. Such a transformation is called a rotation of coordinate axes. It implies not only that F_1, F_2, and F_3 are mutually orthogonal but that, if OX, OY, and OZ is a right-hand system, so also is OX', OY', and OZ'. (See Appendix 2, Section 2A.4, for the definition of a right-hand system.) The first of these conditions is satisfied if $(\lambda_1, \mu_1, \nu_1)$, $(\lambda_2, \mu_2, \nu_2)$, and $(\lambda_3, \mu_3, \nu_3)$ are mutually orthogonal unit vectors. The implications of the second condition will be discussed in Chapter 5 where a more detailed discussion of rotations will be found.

Exercise 1.10

1. Use the method of Section 11 to derive the equations for a transformation of coordinates by rotation of axes in the plane, namely,
$$x = x' \cos \theta - y' \sin \theta,$$
$$y = x' \sin \theta + y' \cos \theta,$$
where θ is the angle through which the axes are rotated.

Vectors and Vector Spaces

2. Find the equations of the ellipse $4x^2 + y^2 = 4$ and of the circle $x^2 + y^2 = 1$ after the coordinate axes have been rotated through angles of (a) $45°$, (b) $30°$.

3. Write the equations for transformation of coordinates in space by a rotation through an angle θ (a) about the z-axis, (b) about the y-axis.

4. Rectangular axes OX, OY, and OZ are rotated through an angle θ about OZ to new positions OX', OY', and OZ' ($OZ' = OZ$). These new axes are then rotated through an angle φ about OX' to new positions OX'', OY'', and OZ'' ($OX'' = OX'$). Find the original coordinates (x, y, z) in terms of the final coordinates (x'', y'', z''). *Hint:* first find x, y, z in terms of x', y', z', then x', y', z', in terms of x'', y'', z'', and then substitute.

5. Verify that in Problem 4
$$x^2 + y^2 + z^2 = x'^2 + y'^2 + z'^2 = x''^2 + y''^2 + z''^2.$$

6. If E_1, E_2, E_3 is a normal orthogonal basis of $\mathscr{V}_3(\mathscr{R})$ and F_1, F_2, F_3 is any other basis, prove that the coordinates (x_1, x_2, x_3) of a vector V relative to the E-basis are given in terms of the coordinates (y_1, y_2, y_3) relative to the F-basis by the equations
$$x_i = \sum_{j=1}^{3} c_{ij} y_j, \qquad (i = 1, 2, 3),$$
where $c_{ij} = E_i \cdot F_j$.

12. VECTORS OVER THE COMPLEX FIELD

In the following chapters we shall often have occasion to use vectors of the form $X = (x_1, x_2, \cdots, x_n)$ whose coordinates $x_1, x_2, \cdots, x_n$ are complex numbers. If it is understood that the coordinates of X may not be real, then X is called a *vector over the complex field* and a scalar is understood to mean any complex number. A set of such vectors which is closed under addition and (complex) scalar multiplication is called a vector space over the field $\mathscr{C}$ of all complex numbers. The set of all vectors of order n over the field of complex numbers is clearly such a vector space. It is denoted by $\mathscr{V}_n(\mathscr{C})$.

A set of vectors of $\mathscr{V}_n(\mathscr{C})$ is linearly dependent if the vectors of the set satisfy a nontrivial linear relation with complex coefficients. If the only such linear relation is the trivial one, they are linearly independent. Since Theorem 1.1 was proved for equations with complex coefficients, it follows as in Theorem 1.3 that any set of more than n vectors of $\mathscr{V}_n(\mathscr{C})$ is linearly dependent. Just as in the real case, this enables us to define the dimension of any subspace $\mathscr{S}$ of $\mathscr{V}_n(\mathscr{C})$ as the maximum number of linearly independent vectors that can be chosen from $\mathscr{S}$.

The definitions and theorems of Sections 7 and 9, concerning subspaces, dimension, generating systems, bases, sums and intersections of spaces, carry through for vector spaces over the complex field without any change other than to substitute $\mathscr{V}_n(\mathscr{C})$ for $\mathscr{V}_n(\mathscr{R})$ and complex scalars for real scalars. These results will be assumed when necessary without further proof. We note also that Theorem 1.2 holds in the complex case; that is, the set of all complex solution vectors of a set of homogeneous linear equations in n unknowns, with complex coefficients, constitutes a vector space $\mathscr{S}$ which is a subspace of $\mathscr{V}_n(\mathscr{C})$. The space $\mathscr{S}$ is called the (complex) solution space of the equations.

The theory of inner products, length, and orthogonality developed in Section 10 leads to some difficulties if applied without modification to vectors over the complex field. For example, if $X = (x_1, x_2, \cdots, x_n)$ is a complex vector, $X \cdot X = x_1^2 + x_2^2 + \cdots + x_n^2$ need no longer be positive so that $\sqrt{X \cdot X}$ is no longer a satisfactory definition for the length of X if we wish the length of every vector to be real. The nonzero vector $(1, i)$, for example, would have zero length and be orthogonal to itself under the definitions of Section 10. These difficulties can be overcome by a simple modification, in the complex case, of the definition of inner product. This will be discussed in Chapter 9.

Exercise 1.11

1. If $X_1, X_2, \cdots, X_r$ is a set of linearly independent real vectors, prove that they are still a linearly independent set when considered as vectors of $\mathscr{V}_n(\mathscr{C})$; that is, if they satisfy no nontrivial linear relation with real coefficients, neither do they satisfy one with complex coefficients.

2. Let $X_1, X_2, \cdots, X_r$ be any r real vectors. Let $\mathscr{S}$ be the vector space over $\mathscr{R}$ consisting of all linear combinations of $X_1, X_2, \cdots, X_r$ with real coefficients, and let $\mathscr{T}$ be the vector space over $\mathscr{C}$ consisting of all such linear combinations with complex coefficients. Use the result of Problem 1 to prove that $\mathscr{S}$ and $\mathscr{T}$ have the same dimension.

It is perhaps well to mention at this stage that the term vector space is usually defined abstractly in such a way as to include not only $\mathscr{V}_n(\mathscr{R})$, $\mathscr{V}_n(\mathscr{C})$, and their subspaces but also spaces of finite or infinite dimension over an arbitrary field of scalars. The student who is interested in this more general approach is referred to Appendix 1, where he will find the abstract definition of a vector space over an arbitrary field, some discussion of its advantages, and some indication of the place that $\mathscr{V}_n(\mathscr{R})$ and $\mathscr{V}_n(\mathscr{C})$ occupy in the more general theory.

CHAPTER TWO

Matrices, Rank, and Systems of Linear Equations

13. BASIC DEFINITIONS

A rectangular array of numbers of the form

$$(1) \quad \begin{pmatrix} a_{11} & a_{12} & \cdots & a_{1n} \\ a_{21} & a_{22} & \cdots & a_{2n} \\ \cdot & \cdot & \cdot & \cdot \\ a_{m1} & a_{m2} & \cdots & a_{mn} \end{pmatrix}$$

is called a *matrix*. The numbers a_{ij} which occur in the array are called the *elements* of the matrix. If all its elements are real numbers, (1) will be called a real matrix, but, if its elements may be complex numbers, it is called a complex matrix, or a matrix over the complex field. We shall assume unless otherwise stated that our matrices have complex elements. The matrix (1) has m *rows* and n *columns*. If the number of rows and columns is understood from the context, (1) will often be written in the abbreviated form (a_{ij}).

Let A be the matrix (1). The elements $a_{11}, a_{22}, a_{33}, \cdots$ are called the elements of the *main diagonal* of A. We can associate with A a second matrix obtained by reflecting the elements of A in the main diagonal; that is, by writing the rows of A as columns. This second matrix is called the transpose of A and is denoted by A^t. We have then

$$A^t = \begin{pmatrix} a_{11} & a_{21} & \cdots & a_{m1} \\ a_{12} & a_{22} & \cdots & a_{m2} \\ \cdot & \cdot & \cdot & \cdot \\ a_{1n} & a_{2n} & \cdots & a_{mn} \end{pmatrix}$$

and A^t has n rows and m columns. A matrix $X = (x_1, x_2, \cdots, x_n)$ consisting of one row and n columns is, of course, an nth order vector. Its transpose,

$$X^t = \begin{pmatrix} x_1 \\ x_2 \\ \cdot \\ \cdot \\ \cdot \\ x_n \end{pmatrix},$$

is a matrix of n rows and one column, and will be called a *column vector* in contrast to X itself, which will be called a *row vector*. The rows of the matrix A will be called the row vectors of A, and the columns of A the column vectors of A. Since it is typographically inconvenient to use the column vector notation, we shall always represent column vectors by the notation X^t, Y^t, etc., where X, Y, etc., are the corresponding row vectors. Some authors reverse this procedure and use X, Y, $\cdots$ for column vectors and X^t, Y^t, $\cdots$, or equivalent notation, for row vectors. There are advantages and disadvantages to both these procedures but there is no unanimity among mathematicians as to which is preferable. The student should therefore be prepared to read books using either notation. We note that, if A is any matrix and A^t its transpose, then A is the transpose of A^t so that $(A^t)^t = A$. A matrix that has the same number of rows as columns is called a *square matrix*. The transpose of a square matrix is also square. A square matrix with n rows and n columns is called a matrix of *order n*.

A matrix is a new kind of mathematical entity, just as were the vectors of order n introduced in Chapter 1. The theory of matrices and its ramifications form an enormous body of mathematical literature, only a small fraction of which will be included in this book. Our first application of matrices will be to the formulation of a complete theory of systems of linear equations. For example, the array of coefficients in equations 7, Section 5, constitute a matrix, and it will be found that the existence of solutions of these equations and the nature of the solutions if they exist are best determined from the properties of the matrix of the coefficients. The theorems of the present chapter will be directed towards this end.

14. RANK OF A MATRIX

Let A be the matrix (1), and suppose that the elements of A are any complex numbers. Let R_1, R_2, $\cdots$, R_m be the row vectors of A and

Matrices, Rank, and Linear Equations

$C_1^t, C_2^t, \cdots, C_n^t$ the column vectors. The vector space $\mathscr{C}(R_1, R_2, \cdots, R_m)$ consisting of all linear combinations of $R_1, R_2, \cdots, R_m$ with complex coefficients will be called the *row space* of A, and $\mathscr{C}(C_1, C_2, \cdots, C_n)$ will be called the *column space* of A.

DEFINITION. The *row rank* of a matrix A is the dimension of its row space, and the *column rank* of A is the dimension of its column space.

In view of Theorem 1.6, the row rank is the maximum number of linearly independent row vectors of A and the column rank is the maximum number of linearly independent column vectors of A.

If A is a real matrix, it is more usual to understand by its row space its *real row space*, namely, the space $\mathscr{R}(R_1, R_2, \cdots, R_m)$ consisting of all linear combinations of $R_1, R_2, \cdots, R_m$ with real coefficients. In view of the result stated in Problem 2, Exercise 1.11, it makes no difference in the definition of the row rank of a real matrix whether we use the real or the complex row space. Similar remarks apply to the column space and column rank of a real matrix.

THEOREM 2.1. *For any matrix A the row rank is equal to the column rank*

Proof. Let A be the matrix (1), and let r be its row rank and c its column rank. Let A' be any matrix obtained from A by permuting the rows of A. Obviously the row rank of A' is the same as that of A. It is easy to see also that A and A' have the same column rank. For, if the column vectors $C_1^t, C_2^t, \cdots, C_n^t$ satisfy a linear relation

$$b_1 C_1^t + b_2 C_2^t + \cdots + b_n C_n^t = O,$$

the coefficients $b_1, b_2, \cdots, b_n$ must be a solution of the system of linear equations

$$\sum_{j=1}^{n} a_{ij} x_j = 0, \qquad (i = 1, 2, \cdots, m)$$

whose coefficient matrix is A. But these same linear equations can be written in a different order so that their coefficient matrix is A', and it then follows that

$$b_1 C_1'^t + b_2 C_2'^t + \cdots + b_n C_n'^t = O,$$

where $C_1'^t, \cdots, C_n'^t$ are the column vectors of A'. Thus every linear relation satisfied by the column vectors of A is also satisfied by the column vectors of A'. Conversely, it follows similarly that every linear relation satisfied by the column vectors of A' is satisfied also by those of A. Hence A and A' have the same column rank.

In view of the above remarks we can assume that the first r rows of A are linearly independent and that the remaining rows are linear combinations of the first r, since such a rearrangement of the rows of A changes neither the row nor column rank. Consider now the matrix B which consists of the first r rows of A. The row rank of B is r since its r rows are linearly independent. We shall show that B has column rank c. The system of equations

(2)
$$a_{11}x_1 + a_{12}x_2 + \cdots + a_{1n}x_n = 0,$$
$$\cdots \cdots \cdots \cdots \cdots$$
$$a_{r1}x_1 + a_{r2}x_2 + \cdots + a_{rn}x_n = 0,$$

whose matrix is B, has precisely the same solutions as the system

(3)
$$a_{11}x_1 + a_{12}x_2 + \cdots + a_{1n}x_n = 0,$$
$$\cdots \cdots \cdots \cdots \cdots$$
$$a_{r1}x_1 + a_{r2}x_2 + \cdots + a_{rn}x_n = 0,$$
$$\cdots \cdots \cdots \cdots \cdots$$
$$a_{m1}x_1 + a_{m2}x_2 + \cdots + a_{mn}x_n = 0,$$

whose matrix is A. For, since the last $m - r$ rows of matrix A are linear combinations of the first r, the last $m - r$ equations of (3) are linear combinations of the first r, and are therefore satisfied by any solution of (2). It follows that the column vectors of B satisfy exactly the same linear relations as the column vectors of A. Hence A and B have the same column rank.

Now the column vectors of B have r coordinates, and hence by Theorem 1.3 not more than r of them can be linearly independent. Hence $c \leq r$, and we have proved that for any matrix the column rank is less than or equal to the row rank. This must be true also, then, of the transposed matrix A^t which has row rank c and column rank r. Hence $r \leq c$, and combining this with the previous result we find $r = c$.

DEFINITION. The *rank* of a matrix A is the common value of its row rank and column rank; that is to say, the maximum number of linearly independent row (or column) vectors of A.

Matrices, Rank, and Linear Equations

The rank of a matrix plays an important role in the theory that follows. A method of finding the rank of any matrix will be given in Chapter 3.

15. DETERMINANTS

In this section the definition and elementary properties of determinants are given for purposes of reference. It is assumed that the reader has some previous acquaintance with these, and the proofs of the simpler properties are therefore omitted. The student who has no knowledge of the subject should read at this point the chapter on determinants from any good textbook on college algebra or theory of equations.†

Let $A = (a_{ij})$ be any square matrix of order n. Consider all possible products of elements of A formed by multiplying together one and only one element from each row and each column. This amounts to forming all products of the form

$$(4) \qquad a_{1i_1} a_{2i_2} \cdots a_{ni_n},$$

where $i_1, i_2, \cdots, i_n$ is some permutation of the numbers $1, 2, \cdots, n$. The fact that every subscript $1, 2, \cdots, n$ occurs exactly once as an initial subscript in the product (4) ensures that exactly one element has been chosen from each row of A. Similarly the fact that each subscript $1, 2, \cdots, n$ occurs exactly once among the final subscripts $i_1, i_2, \cdots, i_n$ ensures that one element has been chosen from each column. The number of products of the form (4) that can be formed is clearly equal to the number of permutations $i_1, i_2, \cdots, i_n$ of the numbers $1, 2, \cdots, n$ in which each number occurs exactly once. This number is $n!$.

Now consider the product (4) in which the initial subscripts occur in natural order. An *inversion* is said to occur in the permutation $i_1, i_2, \cdots, i_n$ of second subscripts whenever a larger subscript precedes a smaller. For example, for $n = 6$, the product

$$a_{13} a_{26} a_{32} a_{41} a_{54} a_{65}$$

has a total of seven inversions of the second subscripts because 3 precedes 1 and 2 (2 inversions), 6 precedes 2, 1, 4, and 5 (4 inversions), and 2 precedes 1 (1 inversion). A permutation $i_1, i_2, \cdots, i_n$ of the numbers $1, 2, \cdots, n$ is said to be *even* or *odd* according as the number of inversions occurring in the permutation is even or odd.

---

† See, for example, F. S. Nowlan, *College Algebra*, McGraw-Hill Book Co., New York, 1947, or L. E. Dickson, *New First Course in the Theory of Equations*, John Wiley & Sons, New York, 1939.

We now define a determinant as follows:

DEFINITION. The *determinant* associated with the square matrix A is denoted by $|A|$ or by

$$\begin{vmatrix} a_{11} & a_{12} & \cdots & a_{1n} \\ \cdot & \cdot & \cdots & \cdot \\ a_{n1} & a_{n2} & \cdots & a_{nn} \end{vmatrix}.$$

It is a polynomial in the elements of A defined by

$$|A| = \Sigma \pm a_{1i_1} a_{2i_2} \cdots a_{ni_n},$$

where the summation extends over all $n!$ permutations $i_1, i_2, \cdots, i_n$ of the subscripts $1, 2, \cdots, n$, and the sign before a term is $+$ or $-$ according as the permutation $i_1, i_2, \cdots, i_n$ is even or odd.

The following properties of determinants follow easily from the definition. Their proofs may be found in the books referred to above. In these theorems A always denotes a square matrix.

D1   The determinant of a matrix A is equal to the determinant of the transposed matrix A^t.

D2   If two rows (or two columns) of the matrix A are interchanged, the determinant of the new matrix so formed is equal to $-|A|$.

D3   If two rows (or two columns) of A are identical, then $|A| = 0$.

D4   If a row vector R of A is replaced by a scalar multiple kR of itself (or a column vector C^t by kC^t), the determinant of the new matrix so formed is equal to $k|A|$.

D5   If a row vector (or a column vector) of A is equal to the zero vector, then $|A| = 0$.

D6   If a scalar multiple of any row vector of A is added to any other row vector of A (or a scalar multiple of any column vector to another column vector), the determinant of the new matrix so formed is equal to $|A|$.

If in any matrix A several rows and/or columns are deleted, the remaining elements form a rectangular array which is called a *submatrix* of A. In particular, if, in a square matrix A, the ith row and jth column (which intersect in the element a_{ij}) are deleted, the remaining submatrix is denoted by A_{ij}. The determinant $|A_{ij}|$ is called the *minor* of the element a_{ij} in A, and $(-1)^{i+j}|A_{ij}|$ is called the *signed minor* or *cofactor* of a_{ij} in A. We can now state the theorem on expansion of determinants by minors as follows:

D7   If $A = (a_{ij})$ is any square matrix, then, for $i, j = 1, 2, \cdots, n$,

$$|A| = (-1)^{i+1}a_{i1}|A_{i1}| + (-1)^{i+2}a_{i2}|A_{i2}| + \cdots + (-1)^{i+n}a_{in}|A_{in}|$$
$$= (-1)^{1+j}a_{1j}|A_{1j}| + (-1)^{2+j}a_{2j}|A_{2j}| + \cdots + (-1)^{n+j}a_{nj}|A_{nj}|.$$

Matrices, Rank, and Linear Equations

From D3 and D7 we see that, if the elements of any row (or column) are multiplied by the cofactors of the corresponding elements of a different row (or column) and the products are added, the result, being the expansion by minors of a determinant with two identical rows (or columns) is zero. This may be stated as follows:

D8 If $A = (a_{ij})$ is a square matrix and if $k \neq i$, then

$$(-1)^{k+1}a_{i1}|A_{k1}| + (-1)^{k+2}a_{i2}|A_{k2}| + \cdots + (-1)^{k+n}a_{in}|A_{kn}| = 0.$$

Similarly, if $k \neq j$,

$$(-1)^{1+k}a_{1j}|A_{1k}| + (-1)^{2+k}a_{2j}|A_{2k}| + \cdots + (-1)^{n+k}a_{nj}|A_{nk}| = 0.$$

The theorems stated in D7 and D8 can be put more compactly by introduction of a useful notational device. The symbol δ_{ij}, known as a *Kronecker delta*, is defined to be equal to one if $i = j$ and equal to zero if $i \neq j$. The range of each subscript is usually clear from the context and in this case is from 1 to n. Now, if we use the notation $|A_{ij}|^*$ for the cofactor of a_{ij} in A, namely, $(-1)^{i+j}|A_{ij}|$, we can write the results of D7 and D8 in the form

(5)
$$\sum_{k=1}^{n} a_{ik}|A_{jk}|^* = |A|\delta_{ij},$$

$$\sum_{i=1}^{n} a_{ij}|A_{ik}|^* = |A|\delta_{jk}.$$

We are now ready to prove what might be called the fundamental theorem on determinants:

THEOREM 2.2. A necessary and sufficient condition that the determinant of a matrix A, of order n, be equal to zero is that the rank of A be less than n; that is, that the row vectors (and also therefore the column vectors) of A are linearly dependent.

Proof. It is easy to see that the condition is sufficient. For, if the rank of A is less than n, the row vectors of A are linearly dependent and hence one row vector, say R_i, is a linear combination of the remaining ones. Hence by subtracting from R_i suitable scalar multiples of the remaining row vectors we can get a new matrix A' whose ith row is the zero vector. By D6 above, $|A| = |A'|$, and, by D5, $|A'| = 0$. The sufficiency of the condition is therefore proved.

Now suppose that $|A| = 0$. We wish to prove that A has rank less than n. The proof will be by induction on the order of the matrix A. The result is clearly true for matrices of order 1. Assume it true for all matrices of order $n - 1$. We divide the proof into two parts:

(a) Suppose that A has at least one minor $|A_{ij}|$ that is not zero. By (5), since $|A| = 0$, we have

$$|A_{i1}|^* a_{k1} + |A_{i2}|^* a_{k2} + \cdots + |A_{in}|^* a_{kn} = 0,$$

for $k = 1, 2, \cdots, n$, and hence

(6) $\qquad |A_{i1}|^* C_1^t + |A_{i2}|^* C_2^t + \cdots + |A_{in}|^* C_n^t = O,$

where $C_1^t, C_2^t, \cdots, C_n^t$ are the column vectors of A. Since at least one coefficient $|A_{ij}|^* \neq 0$ by our assumption, (6) is a nontrivial linear relation among the column vectors of A. Hence these vectors are linearly dependent, and the rank of A is less than n as required.

(b) Suppose that all the minors $|A_{ij}|$ of A are zero. By our induction assumption, since each A_{ij} is a matrix of order $n - 1$ with determinant zero, the rank of each matrix A_{ij} is less than $n - 1$. Consider the matrix B consisting of the first $n - 1$ rows of A. The rank of B must be less than $n - 1$, for, if B had rank $n - 1$, it would have $n - 1$ linearly independent column vectors which together would constitute a square matrix of rank $n - 1$. But this is impossible since this matrix would be one of the $(n - 1)$th-order submatrices A_{ij} of A all of which have rank less than $n - 1$. Hence B has rank less than $n - 1$, and the row vectors of B, and therefore those of A, are linearly dependent. Hence A has rank less than n, and the proof is complete.

DEFINITION. A matrix is said to be *singular* if either its row vectors or column vectors are linearly dependent. If both its row vectors and its column vectors are linearly independent it is said to be *nonsingular*.

It is clear from Theorem 1.3 that a matrix that is not square is always singular, since either the number of row vectors or the number of column vectors must exceed the number of coordinates in these vectors. By Theorem 2.2 a square matrix is singular if and only if its determinant is zero. *A nonsingular matrix is therefore always a square matrix whose rank is equal to its order and whose determinant, therefore, is not zero.*

THEOREM 2.3. A matrix has rank r if and only if it has a nonsingular submatrix of order r but no nonsingular submatrix of order greater than r.

Proof. Let s be the order of the nonsingular submatrix of A of greatest order, so that A has a nonsingular submatrix B of order s but every submatrix of order $s + 1$ is singular. We shall prove that $s = r$, the rank of A. The s rows of A which contain the s rows of B must be linearly independent since, if they are linearly dependent, the row vectors of B would be too, and B would be singular. It follows that A has at least s linearly independent row vectors, and therefore $s \leq r$.

We show next that any $s+1$ rows of A are linearly dependent. Let C be the submatrix of A consisting of any $s+1$ rows of A. The rank of C must be less than $s+1$, for, if C contained $s+1$ linearly independent column vectors, these would constitute a nonsingular submatrix of A of order $s+1$. Hence the $s+1$ rows of C must be linearly dependent. Hence any $s+1$ rows of A are linearly dependent and therefore $r < s+1$, and, since $s \leq r$, we have $s = r$ as required.

In many books the rank of a matrix A is defined as the order of the largest nonsingular submatrix of A. Theorem 2.3 shows that this definition is equivalent to our definition of rank.

Exercise 2.1

1. Evaluate the following determinants:

$$(a) \begin{vmatrix} 2 & -1 & 6 \\ 4 & 1 & 2 \\ 3 & 5 & 7 \end{vmatrix}, \quad (b) \begin{vmatrix} 2 & 7 & 5 & 8 \\ 7 & -1 & 2 & 5 \\ 1 & 0 & 4 & 2 \\ -3 & 6 & -1 & 2 \end{vmatrix}.$$

2. If A is a matrix in which all elements above (or below) the main diagonal are zero, show that $|A|$ is equal to the product of the elements in the main diagonal of A.

3. Using Theorems 2.2 and 2.3, test the following sets of vectors for linear independence:

   (a) $(2, 1, 0)$, $(4, 1, -3)$, $(0, 2, 4)$.

   (b) $(1, 4, 5)$, $(2, -1, -2)$, $(7, 10, 11)$.

   (c) $(1, -1, 2, 2)$, $(3, 5, 2, -4)$, $(5, 3, 6, 1)$.

4. Find x so that the vectors $(2, 1, 1)$, $(1, 3, -2)$, and $(4, 5, x)$ are linearly dependent. For what real values of x are these vectors linearly independent?

5. Find x so that the vectors $(2, 1, 0, 5)$, $(3, 2, 4, 9)$, and $(1, 0, -4, x)$ are linearly dependent.

6. Is it possible to find x so that the vectors $(4, -2, 1, 7)$, $(1, 0, 2, 4)$, and $(2, -1, 5, x)$ are linearly dependent? Why?

7. Find x and y so that the vectors $(4, -2, 1, 7)$, $(1, 0, 2, 4)$, and $(2, -3, x, y)$ are linearly dependent.

8. If every rth-order submatrix of a matrix A is singular, prove that every mth-order submatrix of A, where $m > r$, is also singular.

9. Let $\varphi_{ij}(x)$, $i, j = 1, 2, \cdots, n$, be n^2 differentiable functions of x, and let $\varphi(x)$ be the determinant $|\varphi_{ij}(x)|$. Prove that the derivative $\varphi'(x)$ is equal to the sum of the n determinants $D_1, D_2, \cdots, D_n$, where D_i is the determinant

$|\varphi_{ij}(x)|$ with the functions in its *i*th column (or row) replaced by their derivatives. *Hint:* apply the rule for differentiating a product of *n* functions to each term in the expansion of $|\varphi_{ij}(x)|$, and then group the resulting terms.

10. Let $\varphi_1(x), \cdots, \varphi_n(x)$ be *n* solutions of the linear differential equation, with constant coefficients,

$$\frac{d^n y}{dx^n} + a_1 \frac{d^{n-1}y}{dx^{n-1}} + \cdots + a_n y = 0,$$

and denote by *W* the determinant

$$\begin{vmatrix} \varphi_1(x) & \varphi_2(x) & \cdots & \varphi_n(x) \\ \varphi_1'(x) & \varphi_2'(x) & \cdots & \varphi_n'(x) \\ \cdots & \cdots & \cdots & \cdots \\ \varphi_1^{(n-1)}(x) & \varphi_2^{(n-1)}(x) & \cdots & \varphi_n^{(n-1)}(x) \end{vmatrix},$$

whose *i*th row consists of the $(i-1)$th derivatives of $\varphi_1(x), \cdots, \varphi_n(x)$. Prove that

$$\frac{dW}{dx} = -a_1 W,$$

and hence that $W = W_0 e^{-a_1 x}$, where W_0 is the value of *W* when $x = 0$. (The determinant *W* is called the *Wronskian* of the functions $\varphi_1(x), \cdots, \varphi_n(x)$. The condition $W = 0$, identically in an interval (a, b), is necessary and sufficient for the linear dependence of the solutions $\varphi_1(x), \cdots, \varphi_n(x)$ in that interval. See, for example, R. P. Agnew, *Differential Equations*, McGraw-Hill Book Co., 1942.)

16. SYSTEMS OF LINEAR EQUATIONS

We are now in a position to complete our theory of systems of linear equations.

THEOREM 2.4. A system of *n* linear homogeneous equations in *n* unknowns has a nontrivial solution if and only if the matrix of coefficients is singular; that is, if and only if the determinant of the coefficients is zero.

Proof. Let $A = (a_{ij})$ be the matrix of the coefficients. The equations

$$a_{11}x_1 + a_{12}x_2 + \cdots + a_{1n}x_n = 0,$$
$$\cdots \cdots \cdots \cdots \cdots$$
$$a_{n1}x_1 + a_{n2}x_2 + \cdots + a_{nn}x_n = 0$$

are equivalent to

$$x_1 C_1^t + x_2 C_2^t + \cdots + x_n C_n^t = 0,$$

where $C_1^t, C_2^t, \cdots, C_n^t$ are the column vectors of *A*. They therefore have a nontrivial solution if and only if the column vectors of *A* are linearly

Matrices, Rank, and Linear Equations   49

dependent; that is, if and only if A has rank less than n, or if and only if $|A| = 0$.

THEOREM 2.5. *The solution space of a system of linear homogeneous equations in n unknowns has dimension $n - r$, where r is the rank of the matrix of coefficients.*

Proof. Let the system of equations be

(7)
$$a_{11}x_1 + a_{12}x_2 + \cdots + a_{1n}x_n = 0,$$
$$\cdots \cdots \cdots \cdots \cdots$$
$$a_{m1}x_1 + a_{m2}x_2 + \cdots + a_{mn}x_n = 0.$$

Since the rank of the coefficient matrix is r, only r of these equations are independent, the rest being linear combinations of them. There is no loss of generality in assuming that the first r equations are independent so that the system (7) is equivalent to the system

(8)
$$a_{11}x_1 + a_{12}x_2 + \cdots + a_{1n}x_n = 0,$$
$$\cdots \cdots \cdots \cdots \cdots$$
$$a_{r1}x_1 + a_{r2}x_2 + \cdots + a_{rn}x_n = 0$$

consisting of the first r equations of (7). Since the r rows of the coefficient matrix A of (8) are independent, its rank is r and it therefore has r linearly independent columns. By renaming the unknowns, if necessary, we may assume that the first r columns of the coefficient matrix are independent.

Now, if $r < n$ and we put $x_{r+2} = x_{r+3} = \cdots = x_n = 0$, (8) reduces to a system of r equations in the $r + 1$ unknowns, $x_1, x_2, \cdots, x_{r+1}$. By Theorem 1.1 these have a nontrivial solution. Moreover in this solution $x_{r+1} \neq 0$ since otherwise we would have a nontrivial linear relation among the first r columns of A. Since any scalar multiple of a solution vector of (8) is also a solution vector, it follows that there exists a solution vector of (8) of the form

$$E_1 = (t_{11}, t_{12}, \cdots, t_{1r}, 1, 0, 0, \cdots, 0).$$

In the same manner, putting each time all but one of the unknowns $x_{r+1}, x_{r+2}, \cdots, x_{r+n}$ equal to zero we get solution vectors of (8) of the form

$$E_2 = (t_{21}, t_{22}, \cdots, t_{2r}, 0, 1, 0, \cdots, 0),$$
$$\cdots \cdots \cdots \cdots \cdots$$
$$E_{n-r} = (t_{n-r1}, t_{n-r2}, \cdots, t_{n-rr}, 0, 0, \cdots, 0, 1).$$

Now the solution vectors $E_1, E_2, \cdots, E_{n-r}$ are certainly linearly independent since a linear combination $c_1E_1 + c_2E_2 + \cdots + c_{n-r}E_{n-r}$ has for its last $n-r$ coordinates the coefficients $c_1, c_2, \cdots, c_{n-r}$ and cannot be the zero vector unless $c_1 = c_2 = \cdots = c_{n-r} = 0$. Moreover every solution vector of (8) is a linear combination of $E_1, \cdots, E_{n-r}$. For suppose that $S = (b_1, b_2, \cdots, b_n)$ is any such solution vector. By Theorem 1.2, the vector

$$T = S - b_{r+1}E_1 - b_{r+2}E_2 - \cdots - b_nE_{n-r}$$

is also a solution vector of (8), and therefore of (7), and T has the form $(d_1, d_2, \cdots, d_r, 0, 0, \cdots, 0)$. However, on substituting this solution in (7), we get a linear relation among the first r columns of A with coefficients $d_1, d_2, \cdots, d_r$. Since these columns are linearly independent, we have $d_1 = d_2 = \cdots = d_r = 0$, and therefore $T = O$ and $S = b_{r+1}E_1 + b_{r+2}E_2 + \cdots + b_nE_{n-r}$. Hence $E_1, E_2, \cdots, E_{n-r}$ form a basis for the solution space which therefore has dimension $n - r$.

THEOREM 2.6. *The system of nonhomogeneous equations*

(9) $$\sum_{j=1}^{n} a_{ij}x_j = c_i, \qquad (i = 1, 2, \cdots, m)$$

has a solution if and only if the rank of the matrix $A = (a_{ij})$ of the coefficients is equal to the rank of the "augmented matrix,"

$$A' = \begin{pmatrix} a_{11} & \cdots & a_{1n} & c_1 \\ a_{21} & \cdots & a_{2n} & c_2 \\ \cdot & \cdot & \cdot & \cdot \\ a_{m1} & \cdots & a_{mn} & c_m \end{pmatrix}.$$

Proof. The existence of a solution of these equations implies that the column vector $C^t = (c_1, c_2, \cdots, c_m)^t$ is a linear combination of the column vectors of A, and hence A' has the same number of linearly independent column vectors and therefore the same rank as A. Conversely, if A and A' have the same rank, C^t must be a linear combination of the column vectors of A, and hence the equations have a solution.

THEOREM 2.7. *If S is a fixed solution vector of equations 9 and T is any solution vector of the corresponding homogeneous equations*

(10) $$\sum_{j=1}^{n} a_{ij}x_j = 0 \qquad (i = 1, 2, \cdots, m),$$

Matrices, Rank, and Linear Equations

then $S + T$ is a solution vector of (9). Moreover, as T varies over all solutions of (10), $S + T$ varies over all solutions of (9).

Proof. If $S = (s_1, s_2, \cdots, s_n)$ and $T = (t_1, t_2, \cdots, t_n)$, then

$$\sum_{j=1}^{n} a_{ij}(s_j + t_j) = \sum_{j=1}^{n} a_{ij}s_j + \sum_{j=1}^{n} a_{ij}t_j$$
$$= c_i \qquad (i = 1, 2, \cdots, m),$$

and therefore $S + T$ is a solution vector of (9).

Now suppose that $V = (v_1, v_2, \cdots, v_n)$ is any solution vector of (9). Substituting

$$V - S = (v_1 - s_1, v_2 - s_2, \cdots, v_n - s_n)$$

in (10), we get

$$\sum_{j=1}^{n} a_{ij}(v_j - s_j) = \sum_{j=1}^{n} a_{ij}v_j - \sum_{j=1}^{n} a_{ij}s_j$$
$$= c_i - c_i = 0 \qquad (i = 1, 2, \cdots, m).$$

Hence $V - S$ is a solution vector of (10), say T, and $V = S + T$. Thus every solution vector of (9) has the form $S + T$, where T is a solution vector of (10).

THEOREM 2.8. *A set of n linear nonhomogeneous equations in n unknowns has a unique solution if and only if its coefficient matrix is nonsingular.*

Proof. If the coefficient matrix has rank n, it follows that the augmented matrix also has rank n, and hence, by Theorem 2.6, the system has at least one solution vector S. However, since the coefficient matrix is nonsingular, the only solution vector of the corresponding homogeneous system is, by Theorem 2.4, the zero vector O. Hence, by Theorem 2.7, the only solution vector of the nonhomogeneous system is $S + O$ or S.

Conversely, by Theorem 2.7, if the solution of the nonhomogeneous system is unique, then the corresponding homogeneous system can have only the trivial solution, and therefore the coefficient matrix is nonsingular.

Students who are familiar with the theory of linear differential equations should note the analogy between Theorem 2.7 and the theorem that states that the general solution of a nonhomogeneous linear differential equation is equal to the sum of a particular solution and the general solution of the corresponding homogeneous equation, usually called the complementary function.

Theorems 2.4 through 2.8 might be called the existence theorems for systems of linear equations since they deal with conditions for the existence of solutions and the nature of these solutions when they exist rather than methods of finding solutions. Most methods of solution of systems of linear equations are variations of the method of successive elimination of the unknowns usually learned in elementary algebra. Properties D7 and D8 of determinants, given in Section 15, provide us with a systematic procedure for finding solutions by this method. Consider first a system of n equations in n unknowns of the form

(11)
$$\begin{aligned} a_{11}x_1 + a_{12}x_2 + \cdots + a_{1n}x_n &= c_1, \\ a_{21}x_1 + a_{22}x_2 + \cdots + a_{2n}x_n &= c_2, \\ &\cdots \\ a_{n1}x_1 + a_{n2}x_2 + \cdots + a_{nn}x_n &= c_n. \end{aligned}$$

Using, as before, the notation $|A_{ij}|^*$ for the cofactor of the element a_{ij} in the matrix A of coefficients, we multiply the first of equations 11 by $|A_{11}|^*$, the second by $|A_{21}|^*, \cdots$, the nth by $|A_{n1}|^*$ and add the resulting equations. By equations 5 the result is

(12) $\qquad |A|x_1 = c_1|A_{11}|^* + c_2|A_{21}|^* + \cdots + c_n|A_{n1}|^*.$

The right-hand side of equation 12 is the expansion by elements of the first column of the determinant

$$A_1 = \begin{vmatrix} c_1 & a_{12} & \cdots & a_{1n} \\ c_2 & a_{22} & \cdots & a_{2n} \\ \cdot & \cdot & & \cdot \\ c_n & a_{n2} & \cdots & a_{nn} \end{vmatrix}.$$

Similarly, using as multipliers for equations 11 the cofactors of the elements of the second, third, $\cdots$, nth columns of A, we find

(13) $\qquad |A|x_i = |A_i| \qquad (i = 1, 2, \cdots, n),$

where A_i is the matrix A with its ith column replaced by the column vector $(c_1, c_2, \cdots, c_n)^t$ of the constant terms of (11). Now, if $|A| \neq 0$, equations 13, on division by $|A|$, provide us with a solution of (11) because in this case there exists, by Theorem 2.8, exactly one solution of (11) and every solution of (11) is certainly a solution of (13).

Now consider an arbitrary system

(14) $\qquad \sum_{j=1}^{n} a_{ij}x_j = c_i \qquad (i = 1, 2, \cdots, m)$

Matrices, Rank, and Linear Equations 53

of m equations in n unknowns. If the ranks of the coefficient matrix A and the augmented matrix A' are not equal, there are no solutions. (Methods of determining these ranks will be given in Chapter 3.) Suppose then that A and A' have the same rank r. By Theorem 1.20, A has a nonsingular submatrix of order r. By reordering the equations and renaming the unknowns if necessary, it may be assumed that the r-by-r submatrix in the upper left-hand corner of A is nonsingular. Since the matrix B consisting of the first r rows of A' also has this nonsingular submatrix, it follows again from Theorem 1.20 that B has rank r. Since A' also has rank r the last $m - r$ rows of A' are linear combinations of the first r, and hence the last $m - r$ of equations 14 are linear combinations of the first r. Hence any solution of these first r equations is a solution of the whole system (14). If now, in the first r equations of (14), we assign arbitrary values to $x_{r+1}, x_{r+2}, \cdots, x_n$, we get a system of r equations in $x_1, x_2, \cdots, x_r$ whose coefficient matrix is nonsingular and which therefore, by Theorem 2.8, has a unique solution. These values of $x_1, x_2, \cdots, x_r$ together with those assigned to $x_{r+1}, \cdots, x_n$ give a solution of the first r equations and therefore of (14). We therefore get one solution of (14) for every choice of values for $x_{r+1}, \cdots, x_n$. In practice the method of determinants described above can be used to solve the first r equations of (14) for $x_1, x_2, \cdots, x_r$ in terms of $x_{r+1}, \cdots, x_n$, and all solutions of (14) are then obtained by assigning all possible values to $x_{r+1}, \cdots, x_n$. This method applies also when equations 14 are homogeneous.

The student should perhaps be warned that the method described above for solving linear equations by determinants (usually known as Cramer's rule) is really more of theoretical than of practical value. If the number of equations is large, say 10 or more, the labor involved in evaluating the necessary determinants is usually prohibitive for a human computer, and for $n > 25$ it is prohibitive even for an electronic computer that can perform 2600 multiplications per second. Many ingenious methods of successive approximation to solutions of such linear systems have been devised. The student who is interested in these is referred to [4]† for a survey and bibliography of such methods.

Exercise 2.2

1. Find all solutions of the following systems of equations:

(a) $\quad x + y - z = 6,$
$\quad\quad 2x + 5y - 2z = 10.$

(b) $\quad x + y = 0,$
$\quad\quad 2x + 3y = 0.$

† Numbers in square brackets refer to references at the back of the book.

(c) $\quad x + y + z = 1,$
$\quad 2x - 3y + 7z = 0,$
$\quad 3x - 2y + 8z = 4.$

(d) $\quad x + y - 2z + t = 0,$
$\quad 2x + 2y - 5z + 3t = 0.$

(e) $\quad x - y + 2z = 1,$
$\quad x + y + z = 2,$
$\quad 2x - y + z = 5.$

(f) $\quad x - y + 2z = 4,$
$\quad 3x + y + 4z = 6,$
$\quad x + y + z = 1.$

2. If $X_1, X_2, \cdots, X_m$ are linearly independent vectors, prove that the vectors $Y_1, Y_2, \cdots, Y_m$, where

$$Y_i = \sum_{j=1}^{m} a_{ij} X_j$$

are linearly independent if and only if the matrix $A = (a_{ij})$ is nonsingular.

3. Find a basis for the solution space of each of the following systems:

(a) $\quad x + 2y - x + t - 2u = 0,$
$\quad 2x + 5y - 3z - t + u = 0.$

(b) $\quad x - 2y + z = 0,$
$\quad y - z + t = 0,$
$\quad z - 2t + u = 0.$

4. Find all solutions of the system
$$x + y - 2z + t = 4,$$
$$2x + 3y + z - t = 10$$

in terms of the appropriate number of arbitrary parameters. Check your result by substitution in the equations.

5. Show that Cramer's rule can be applied directly to solve the equations

$$\sum_{j=1}^{n} a_{ij} X_j = C_i \qquad (i = 1, 2, \cdots, n)$$

for the vectors $X_1, \cdots, X_n$ in terms of the vectors $C_1, \cdots, C_n$.

6. Let $E_1 = (1, 0, 0,)$, $E_2 = (0, 1, 0)$, $E_3 = (0, 0, 1)$. If
$$2E_1 + 5E_2 - 3E_3 = F_1,$$
$$E_1 - E_2 + 2E_3 = F_2$$
$$3E_1 - 2E_2 + E_3 = F_3,$$

use the method suggested in Problem 5 to solve for E_1, E_2, E_3 in terms of F_1, F_2, F_3. Check your result by substitution.

7. Express the vectors $E_1 = (1, 0, 0)$, $E_2 = (0, 1, 0)$, and $E_3 = (0, 0, 1)$ as linear combinations of the vectors $F_1 = (1, 2, 4)$, $F_2 = (-2, 1, 5)$, and $F_3 = (-1, -1, 2)$.

8. Express the vectors $X_1 = (2, 1, 0)$, $X_2 = (3, 1, 1)$, and $X_3 = (2, -1, 7)$ as linear combinations of $Y_1 = (0, 2, 1)$, $Y_2 = (-1, 6, 2)$, and $Y_3 = (1, 3, 4)$,

and then solve for Y_1, Y_2, Y_3 and terms of X_1, X_2, X_3. *Hint:* express X_1, X_2, X_3 in terms of E_1, E_2, E_3 and then E_1, E_2, E_3 in terms of Y_1, Y_2, Y_3, as in Problem 7.

9. If $E_1, E_2, \cdots, E_m$ is a basis of a space $\mathscr{S}$, show that $F_1, F_2, \cdots, F_m$ is a basis of $\mathscr{S}$ if and only if

$$F_i = \sum_{j=1}^{n} a_{ij} E_j \qquad (i = 1, 2, \cdots, m),$$

where (a_{ij}) is nonsingular.

CHAPTER THREE

Further Algebra of Matrices

17. MULTIPLICATION OF MATRICES

Suppose n variables $x_1, x_2, \cdots, x_n$ are defined in terms of m other variables $y_1, y_2, \cdots, y_m$ by a set of linear equations of the form

(1)
$$x_1 = a_{11}y_1 + a_{12}y_2 + \cdots + a_{1m}y_m,$$
$$\cdots \cdots \cdots \cdots \cdots$$
$$x_n = a_{n1}y_1 + a_{n2}y_2 + \cdots + a_{nm}y_m,$$

and $y_1, y_2, \cdots, y_m$ are defined in terms of r new variables $z_1, z_2, \cdots, z_r$ by the equations

(2)
$$y_1 = b_{11}z_1 + b_{12}z_2 + \cdots + b_{1r}z_r,$$
$$\cdots \cdots \cdots \cdots \cdots$$
$$y_m = b_{m1}z_1 + b_{m2}z_2 + \cdots + b_{mr}z_r.$$

It is possible, by substituting the values for $y_1, \cdots, y_m$ from (2) in equations 1, to find a set of linear equations that define the x's in terms of the z's. Suppose that the equations so obtained are

(3)
$$x_1 = c_{11}z_1 + c_{12}z_2 + \cdots + c_{1r}z_r,$$
$$\cdots \cdots \cdots \cdots \cdots$$
$$x_n = c_{n1}z_1 + c_{n2}z_2 + \cdots + c_{nr}z_r.$$

Algebra of Matrices

Let $A = (a_{ij})$ be the matrix of coefficients in (1), $B = (b_{ij})$ be that of (2), and $C = (c_{ij})$ be that of (3). By actually carrying out the substitution of (2) in (1), we find that the element in the ith row and jth column of C is

(4) $$c_{ij} = \sum_{k=1}^{m} a_{ik} b_{kj}.$$

DEFINITION. *The matrix $C = (c_{ij})$ whose elements are defined by (4) is called the product of the matrices A and B in the order named.* We write $C = AB$.

When the matrices A and B have real elements, the rule for forming the product of A and B can be formulated as follows:

The element in the ith row and jth column of the product AB is the inner product of the ith-row vector of A and the jth-column vector of B.

The rule for multiplying complex matrices is, of course, the same as for real matrices but it cannot be stated in terms of inner products of row and column vectors since these have not been defined for complex vectors.

A product AB of two matrices is defined only if the number of columns in A is equal to the number of rows in B. If A has n rows and m columns, we say it is of type (n, m). If B is of type (m, r) then the product AB is of type (n, r). Whenever matrix products are used in what follows it will be assumed, even though not stated explicitly, that the matrices are of such type that the indicated products are defined. It should be noted that the product BA is in general different from AB. In fact, BA is not even defined unless $r = n$. If A and B are square matrices of order n, then both AB and BA are defined and are square matrices of order n. However, in general they are not equal. For example,

$$\begin{pmatrix} 1 & 2 \\ 2 & -1 \end{pmatrix} \begin{pmatrix} 2 & 1 \\ 0 & 4 \end{pmatrix} = \begin{pmatrix} 2 & 9 \\ 4 & -2 \end{pmatrix},$$

but

$$\begin{pmatrix} 2 & 1 \\ 0 & 4 \end{pmatrix} \begin{pmatrix} 1 & 2 \\ 2 & -1 \end{pmatrix} = \begin{pmatrix} 4 & 3 \\ 8 & -4 \end{pmatrix}.$$

If we multiply the matrix AB on the right by a matrix $D = (d_{ij})$ of type (r, s), we find

$$(AB)D = (c_{ij})(d_{ij}) = \left(\sum_{k=1}^{m} a_{ik} b_{kj} \right)(d_{ij}) = \left(\sum_{h=1}^{r} \sum_{k=1}^{m} a_{ik} b_{kh} d_{hj} \right) = (e_{ij}).$$

Now, if we multiply the three matrices A, B, D as follows,

$$A(BD) = (a_{ij})\left(\sum_{h=1}^{r} b_{ih}d_{hj}\right) = \left(\sum_{k=1}^{m}\sum_{h=1}^{r} a_{ik}b_{kh}d_{hj}\right) = (f_{ij}),$$

we find that $(e_{ij}) = (f_{ij})$ since the two double sums are identical except for the order in which the terms are added. If the student has difficulty in seeing this, he should write out the two double sums in full for the case $n = m = r = s = 2$. Hence we have shown:

Matrix multiplication is associative; that is to say, if A, B, D, are any three matrices of types (n, m), (m, r) and (r, s), respectively, then $(AB)D = A(BD)$.

The associativity of matrix multiplication is, of course, implicit in the definition. For, if

$$x_i = \sum_{j=1}^{m} a_{ij}y_j, \quad y_i = \sum_{j=1}^{r} b_{ij}z_j \quad \text{and} \quad z_i = \sum_{j=1}^{s} d_{ij}t_j,$$

then both $(AB)D$ and $A(BD)$ represent the matrix of coefficients of the equations giving the x's in terms of the t's; the two different forms arising only by virtue of the order in which the intermediate substitutions are made. Since this order does not affect the final equations, $(AB)D = A(BD)$.

We shall now prove two theorems concerning matrix products that we shall require in what follows.

THEOREM 3.1. *The transpose of a product of two matrices is equal to the product of their transposes in reverse order, or, in symbols, if A and B are any two matrices whose product AB is defined, then $(AB)^t = B^t A^t$.*

Proof. If $A = (a_{ij})$ and $B = (b_{ij})$ are matrices of type (m, n) and (n, r) respectively, then the element in the ith row and jth column of AB is

$$\sum_{k=1}^{n} a_{ik}b_{kj}.$$

Now

$$B^t = \begin{pmatrix} b_{11} & b_{21} & \cdots & b_{n1} \\ b_{12} & b_{22} & \cdots & b_{n2} \\ \cdot & \cdot & \cdot & \cdot \\ b_{1r} & b_{2r} & \cdots & b_{nr} \end{pmatrix}, \qquad A^t = \begin{pmatrix} a_{11} & a_{21} & \cdots & a_{m1} \\ a_{12} & a_{22} & \cdots & a_{m2} \\ \cdot & \cdot & \cdot & \cdot \\ a_{1n} & a_{2n} & \cdots & a_{mn} \end{pmatrix},$$

Algebra of Matrices

and therefore $B^t A^t$ is defined, and the element in the ith row and jth column of this product is

$$\sum_{k=1}^{n} b_{ki} a_{jk} = \sum_{k=1}^{n} a_{jk} b_{ki}.$$

Since this is the element in the jth row and ith column of AB, it follows that $B^t A^t = (AB)^t$ as required.

THEOREM 3.2. *The rank of the product of two matrices is less than or equal to the rank of either factor.*

Proof. It follows from the definition of a matrix product that the row vectors of AB are linear combinations of the row vectors of B. For example, if $A = (a_{ij})$ and the row vectors of B are $B_1, B_2, \cdots, B_n$, then the first row vector of AB is $a_{11} B_1 + a_{12} B_2 + \cdots + a_{1n} B_n$. Hence the row space of AB is a subspace of the row space of B and therefore has dimension less than or equal to that of the row space of B. Thus the rank of AB is less than or equal to that of B, the second factor in the product.

Now consider $(AB)^t = B^t A^t$. By the above result

$$\text{rank } (AB)^t \leq \text{rank } A^t.$$

But, since the rank of a transposed matrix is equal to that of the original matrix, we have

$$\text{rank } (AB) \leq \text{rank } A.$$

The proof is therefore complete.

Exercise 3.1

1. Given

$$A = \begin{pmatrix} 2 & 0 & -1 \\ 1 & 5 & 2 \\ -2 & 1 & 1 \end{pmatrix}, \quad B = \begin{pmatrix} 2 & 1 & 4 \\ -1 & -1 & 2 \\ 3 & 0 & 1 \end{pmatrix},$$

find A^2, AB, BA, and B^2.

2. Verify by actual multiplication that $(AB)C = A(BC)$, if

$$A = \begin{pmatrix} a_{11} & a_{12} \\ a_{21} & a_{22} \end{pmatrix}, \quad B = \begin{pmatrix} b_{11} & b_{12} \\ b_{21} & b_{22} \end{pmatrix}, \quad C = \begin{pmatrix} c_{11} & c_{12} \\ c_{21} & c_{22} \end{pmatrix}.$$

3. If $x_1 = 2y_1 - y_2 + 3y_3$, $x_2 = y_1 + 2y_2 - y_3$ and $y_1 = z_1 + z_2$, $y_2 = z_1 - z_2$, $y_3 = 2z_1 + z_2$, use matrix multiplication to write the equations giving x_1 and x_2 in terms of z_1 and z_2.

4. If A and B are square matrices of order n in which all the elements above the main diagonal are 0, show that AB has the same property.

5. If $X = (2, -1, 4, 6)$ and $Y = (1, 0, -2, 3)$, write down the matrices XY^t and X^tY.

6. If A and B are the matrices given in Problem 1, show that $|AB| = |A| \, |B|$.

7. Prove that, for any second-order matrices A and B, $|AB| = |A| \, |B|$. (This result will be proved in general in Section 22.)

8. Deduce from Theorem 3.1 that, if $A = B_1 B_2 \cdots B_r$, then

$$A^t = B_r^t \cdots B_2^t B_1^t.$$

9. A matrix A is said to be *symmetric* if $A = A^t$. Prove (a) if P is any matrix, PP^t is symmetric; (b) if A and B are symmetric, AB is symmetric if and only if $AB = BA$.

10. If

$$P = \begin{pmatrix} 2 & -1 & 4 & 0 \\ 3 & 2 & -1 & 4 \end{pmatrix},$$

compute PP^t and P^tP.

18. INVERSES AND ZERO DIVISORS

If δ_{ij} $(i, j = 1, 2, \cdots, n)$ is a Kronecker delta, the square matrix (δ_{ij}) has all its main diagonal elements equal to one and all other elements equal to zero. This matrix is called the *unit matrix of order n* and will be denoted by I_n, or, when its order is clear from the context, simply by I. The unit matrix (also called the *identity matrix*) has the property that $AI = A$ and $IB = B$, where A and B are any two matrices for which the indicated products are defined. In particular, if A is a square matrix of order n, then $AI = IA = A$.

A *zero matrix* is one all the elements of which are equal to zero. A zero matrix will be denoted by O, regardless of its type. We then have, for any matrix A,

$$AO = OA = O,$$

it being understood that when a zero matrix occurs as a factor in a product it is of such type that the indicated product is defined. Thus, if A is of type (m, n), the three O's in the above equation would be of types (n, n), (m, m), and (m, n), respectively. It should be noted that, whereas a unit matrix is always square, a zero matrix need not be.

DEFINITION. An *inverse* of a matrix A is a matrix, denoted by A^{-1}, having the property that both $A^{-1}A$ and AA^{-1} are unit matrices.

Algebra of Matrices

THEOREM 3.3. A matrix A has an inverse if and only if it is nonsingular. A nonsingular matrix has only one inverse.

Proof. Suppose that A is of type (m, n) and that $m \leq n$. If A has an inverse A^{-1}, the products AA^{-1} and $A^{-1}A$ are both defined, and hence A^{-1} must be of type (n, m). It follows that $A^{-1}A = I_n$. If A has rank r, we then have $r \leq m \leq n \leq r$, where the last inequality follows from Theorem 3.2, since the rank of I_n is n. Hence $m = n = r$, and A is nonsingular. If $n \leq m$, the result follows similarly, using $AA^{-1} = I_m$.

Now suppose that $A = (a_{ij})$ is nonsingular. Since $|A| \neq 0$, we can let

$$b_{ij} = \frac{|A_{ji}|^*}{|A|} \qquad (i, j = 1, 2, \cdots, n),$$

where $|A_{ij}|^*$ is the cofactor of a_{ij} in A. Let B be the matrix (b_{ij}). By equations 5, Section 15, we have

$$AB = \left(\sum_{k=1}^{n} a_{ik} b_{kj}\right) = \left(\sum_{k=1}^{n} \frac{a_{ik}|A_{jk}|^*}{|A|}\right) = \left(\frac{|A|\delta_{ij}}{|A|}\right) = I.$$

Similarly,

$$BA = \left(\sum_{k=1}^{n} b_{ik} a_{kj}\right) = \left(\sum_{k=1}^{n} a_{kj} \frac{|A_{ki}|^*}{|A|}\right) = I.$$

Hence B is an inverse of A, and every nonsingular matrix has an inverse.

Finally, to see that the inverse of a nonsingular matrix is unique, suppose that A is nonsingular and A^{-1} is its inverse as determined above. If B is another inverse of A, we have $AB = I$. Multiplying each side of this equation on the left by A^{-1}, we get $A^{-1}(AB) = A^{-1}$ or $A^{-1} = (A^{-1}A)B = IB = B$. Hence A has only one inverse A^{-1}, and this can be found as above.

THEOREM 3.4. If A is nonsingular, the rank of AB is equal to the rank of B and the rank of CA is equal to that of C.

Proof. By Theorem 3.2, rank $AB \leq$ rank B. But, since A is nonsingular, it has an inverse A^{-1} and $B = A^{-1}(AB)$. Applying Theorem 3.2 again, we see that rank $B \leq$ rank AB, and the result follows. The proof for CA is similar.

COROLLARY. The product of two nonsingular matrices is nonsingular.

THEOREM 3.5. If A and B are nonsingular matrices, $(AB)^{-1} = B^{-1}A^{-1}$. Also $(A^{-1})^{-1} = A$, and $(A^{-1})^t = (A^t)^{-1}$.

Proof. Since $(AB)(B^{-1}A^{-1}) = A(BB^{-1})A^{-1} = AA^{-1} = I$, the first result follows.

62 Linear Algebra for Undergraduates

Similarly $A^{-1}A = AA^{-1} = I$ implies $(A^{-1})^{-1} = A$, and $A^t(A^{-1})^t = (A^{-1}A)^t = I^t = I$ shows that $(A^{-1})^t = (A^t)^{-1}$.

COROLLARY. If $A_1, A_2, \cdots, A_r$ are nonsingular matrices of the same order, their product is nonsingular and

$$(A_1 A_2 \cdots A_r)^{-1} = A_r^{-1} \cdots A_2^{-1} A_1^{-1}.$$

If n is any positive integer and A is a square matrix, the notation A^n is used, just as in the case of numbers, to mean the product $AA \cdots A$ of n factors A. It then follows from the above corollary that, if A is nonsingular, $(A^n)^{-1} = (A^{-1})^n$. It is therefore natural to define negative and zero powers of a nonsingular matrix by

$$A^{-n} = (A^{-1})^n \quad \text{and} \quad A^0 = I.$$

It follows that, for any nonsingular matrix A and any integers m and n,

$$A^m A^n = A^{m+n}$$

and

$$(A^m)^n = A^{mn}.$$

The fact that not every matrix has an inverse or "reciprocal" is in a sense responsible for another peculiarity of matrix algebra, namely, the fact that the product of two nonzero matrices may be O. The reader may verify, for example, the equation

$$\begin{pmatrix} 1 & -2 \\ 3 & -6 \end{pmatrix} \begin{pmatrix} 4 & 2 \\ 2 & 1 \end{pmatrix} = \begin{pmatrix} 0 & 0 \\ 0 & 0 \end{pmatrix}.$$

DEFINITION. A nonzero matrix A is called a *zero divisor* if there exists either a nonzero matrix B such that $AB = O$ or a nonzero matrix C such that $CA = O$.

It is clear that a matrix with an inverse could not be a zero divisor. For, if $AB = O$ or $CA = O$ and A has an inverse A^{-1}, it follows that $A^{-1}AB = B = O$ or $CAA^{-1} = C = O$, and hence A cannot be a zero divisor. It therefore follows from Theorem 3.3 that all zero divisors are singular. The following theorem will show that the converse of this statement also holds.

THEOREM 3.6. A matrix is a zero divisor if and only if it is singular.
Proof. Let $A = (a_{ij})$ be any singular matrix of type (m, n) and suppose

first that $m \leq n$. It follows, by Theorem 1.1, if $m < n$ or by Theorem 2.4, if $m = n$, that the equations

(5)
$$a_{11}x_1 + a_{12}x_2 + \cdots + a_{1n}x_n = 0,$$
$$\cdots \cdots \cdots \cdots$$
$$a_{m1}x_1 + a_{m2}x_2 + \cdots + a_{mn}x_n = 0$$

have nontrivial solutions. Let B be any matrix of type (n, r) all the column vectors of which are nonzero solution vectors of (5). Then $B \neq O$ and $AB = O$, so that A is a zero divisor.

Now, if $n < m$, we consider the matrix A^t for which the number of rows is less than the number of columns. Hence the above argument applies, and there exists a nonzero matrix C^t such that $A^t C^t = O$. Hence $(CA)^t = O$ and $CA = O$, whence A is again a zero divisor.

The converse part of the theorem, that, if A is a zero divisor, it must be singular has already been proved above. It also follows from the fact that the matrix equation $AB = O$ implies that every column vector of B is a solution vector of (5). If A is nonsingular, these equations have only the trivial solution, and therefore $B = O$. A similar argument, using transposes, shows that, if $CA = O$ and A is nonsingular, then $C = O$.

Exercise 3.2

1. If

$$A = \begin{pmatrix} 1 & -3 & 2 \\ 4 & 1 & -1 \\ -3 & 2 & 5 \end{pmatrix} \quad \text{and} \quad B = \begin{pmatrix} 2 & 1 & 5 \\ -1 & -2 & -2 \\ 3 & 1 & 2 \end{pmatrix},$$

find the matrices AB, BA, A^{-1}, B^{-1}, $(AB)^{-1}$. Check your results by verifying the relations $(AB)^{-1} = B^{-1}A^{-1}$, $AA^{-1} = I$, $B^{-1}B = I$.

2. Let $A = (a_{ij})$ be any nth-order square matrix, and let A' be the matrix $(|A_{ji}|^*)$ in which the element in the jth row and ith column is the cofactor of a_{ij} in A. Prove that $AA' = |A|(\delta_{ij})$.

3. Use the result of Problem 2 to prove that, if A is a singular matrix of order n, the column vectors of A' are solution vectors of the system of linear homogeneous equations whose coefficient matrix is A.

4. Use the result of Problem 3 to derive the well-known rule that a solution of the equations

$$a_1 x + b_1 y + c_1 z = 0,$$
$$a_2 x + b_2 y + c_2 z = 0$$

is given by

$$x:y:z = \begin{vmatrix} b_1 & c_1 \\ b_2 & c_2 \end{vmatrix} : -\begin{vmatrix} a_1 & c_1 \\ a_2 & c_2 \end{vmatrix} : \begin{vmatrix} a_1 & b_1 \\ a_2 & b_2 \end{vmatrix}.$$

(*Hint:* consider a third equation all the coefficients of which are zero.)

5. Generalize Problem 4 to give a rule for writing down a solution of any $n-1$ independent linear homogeneous equations in n unknowns.

6. Using the method indicated in Problems 4 and 5, solve the following systems of equations:

(a) $3x - y + z = 0$,
   $2x + y - 2z = 0$.

(b) $x - 2y + z + t = 0$,
    $2x + y - z + 2t = 0$,
    $x + y + 2z - t = 0$.

(c) $x + 2z - t = 0$,
    $y + z - 3t = 0$,
    $x + t = 0$.

7. Given

$$A = \begin{pmatrix} 2 & -5 & 3 \\ 1 & 4 & -1 \\ 7 & 2 & 3 \end{pmatrix},$$

prove that A is singular, and find nonzero third-order matrices B and C such that $AB = 0$ and $CA = 0$.

8. If A is of order n and rank r and B is any matrix such that $AB = O$, prove that the rank of B is less than or equal to $n - r$. Prove also that there exists such a matrix B whose rank is equal to $n - r$.

9. If

$$A = \begin{pmatrix} 1 & 1 & 2 \\ 2 & 2 & 4 \\ 3 & 3 & 6 \end{pmatrix},$$

find a matrix B of rank 2 such that $AB = O$.

10. Prove that the row vectors of the matrix product AB are $R_1B, R_2B, \cdots, R_nB$, where $R_1, R_2, \cdots, R_n$ are the row vectors of A.

11. Prove that the column vectors of the matrix product AB are $AC_1^t, AC_2^t, \cdots, AC_n^t$, where $C_1^t, C_2^t, \cdots, C_n^t$ are the column vectors of B.

12. If $A = (a_{ij})$ is nonsingular and

$$\sum_{j=1}^{n} a_{ij}x_j = y_i \qquad (i = 1, \cdots, n,)$$

show that

$$\sum_{i=1}^{n} b_{ij}y_i = x_j,$$

where

$$(b_{ij}) = A^{-1}.$$

Algebra of Matrices

19. ADDITION OF MATRICES AND THE LAWS OF MATRIX ALGEBRA

If $A = (a_{ij})$ and $B = (b_{ij})$ are two matrices of the same type (m, n), their *sum* $A + B$ is defined to be the matrix $(a_{ij} + b_{ij})$ obtained by adding corresponding elements of A and B. It is clear that addition of matrices is commutative, $A + B = B + A$, and associative, $(A + B) + C = A + (B + C)$. Moreover the zero matrix O satisfies the laws $A + O = O + A = A$ for every matrix A, and to every matrix $A = (a_{ij})$ there corresponds a *negative* $-A = (-a_{ij})$ having the property that $A + (-A) = O$.

Finally we shall prove that matrix multiplication is distributive with respect to matrix addition; that is, if A and B are of type (m, n), C of type (r, m), and D of type (n, s), then

$$C(A + B) = CA + CB$$

and

$$(A + B)D = AD + BD.$$

Suppose that the elements in the ith row and jth column of A, B, and C are, respectively, a_{ij}, b_{ij}, and c_{ij}. Then the element in the ith row and jth column of $A + B$ is $a_{ij} + b_{ij}$ and that of $C(A + B)$ is

$$\sum_{k=1}^{m} c_{ik}(a_{kj} + b_{kj}).$$

This is equal to

$$\sum_{k=1}^{m} c_{ik}a_{kj} + \sum_{k=1}^{m} c_{ik}b_{kj}$$

which is the element in the ith row and jth column of $CA + CB$. This proves the first distributive law. The proof of the second is similar and is left as an exercise for the student.

It is sometimes useful also to define a *scalar product* of a number and a matrix. If $A = (a_{ij})$ is any matrix and k a complex number, the product kA is defined to be the matrix (ka_{ij}); that is, to multiply A by k we multiply each of its elements by k. This definition conforms with scalar multiplication of vectors which is a special case of it. It obeys the same algebraic laws as scalar multiplication in a vector space.

If we denote by $\mathscr{S}$ the set of all square matrices of order n with complex elements, we can list the laws of algebra in $\mathscr{S}$ as follows:

Addition

(1) Any two matrices A and B in $\mathscr{S}$ have a sum $A + B$ which belongs to $\mathscr{S}$.
(2) Addition is commutative and associative.

(3) $\mathscr{S}$ contains a zero matrix O such that $A + O = O + A = A$ for every A in $\mathscr{S}$.
(4) Every matrix A of $\mathscr{S}$ has a negative $-A$ in $\mathscr{S}$ such that $A + (-A) = O$. If A and B belong to $\mathscr{S}$, the difference $A - B$ is defined to be the matrix X such that $B + X = A$. It follows that $A - B = A + (-B)$.

Multiplication

(1) Any two matrices A and B in $\mathscr{S}$ have a product AB which belongs to $\mathscr{S}$.
(2) Multiplication is associative: $(AB)C = A(BC)$.
(3) There is a unit matrix I in $\mathscr{S}$ such that $AI = IA = A$ for every A in $\mathscr{S}$.
(4) Multiplication is distributive with respect to addition:

$$A(B + C) = AB + AC,$$
$$(B + C)A = BA + CA.$$

A set $\mathscr{S}$ in which addition and multiplication are defined and satisfy the above laws is called a *ring*. The set of all nth-order square matrices therefore constitutes a ring. A ring is a more general algebraic system than a field since the quotient of two elements of a ring need not be defined. Every field is a ring but the converse is not true. Every "number" in a field except zero must have an inverse, or reciprocal, in the field. This is not necessary in a ring, and in fact we have seen that it is not true in the above ring of matrices, since the singular matrices have no inverses. The general definition of a ring does not require that it contain a unit. Our ring of all square matrices of order n is therefore called a ring with unit element.

20. SOME NOTATIONAL ADVANTAGES OF MATRIX PRODUCTS

The definition of matrix products provides some very useful simplifications of notation. We note first that, if X and Y are nth-order row vectors and if Y^t is the transpose of Y, that is, the column vector with the same coordinates as Y, then the matrix product XY^t is defined and will be a matrix of one row and one column. Such a matrix is not distinguished from the number that is its only element, so that, if $X = (x_1, x_2, \cdots, x_n)$ and $Y = (y_1, y_2, \cdots, y_n)$, we write

$$XY^t = x_1y_1 + x_2y_2 + \cdots + x_ny_n.$$

If X and Y are real vectors, this is simply the inner product $X \cdot Y$. Thus the inner product of two real vectors X and Y may be identified with the matrix product XY^t of X and the transpose of Y.

The expression
$$f = \sum_{i=1}^{n} \sum_{j=1}^{n} a_{ij} x_i y_j$$
is called a *bilinear form* in the $2n$ variables $x_1, x_2, \cdots, x_n$ and $y_1, y_2, \cdots, y_n$. It consists of n^2 terms all of which are of the first degree in the x's and of the first degree in the y's. The n^2 coefficients a_{ij} constitute a square matrix (a_{ij}) of order n, which is called the *matrix of the bilinear form*. If $X = (x_1, x_2, \cdots, x_n)$ and $Y = (y_1, y_2, \cdots, y_n)$ and $A = (a_{ij})$, the reader may easily verify that the matrix product XAY^t is defined, that it is of type (1, 1), and that its one element is equal to the bilinear form f. Hence every bilinear form may be written as a matrix product XAY^t, where A is the matrix of the form and X and Y are the vectors defined above. In the particular case in which $A = I$, the bilinear form f reduces to $XY^t = x_1 y_1 + \cdots + x_n y_n$, so that in the case of two real vectors the inner product $X \cdot Y$ is a special case of a bilinear form in the coordinates of X and Y.

If in the bilinear form f we put $Y = X$, it reduces to a *quadratic form* in $x_1, x_2, \cdots, x_n$, that is to say, a homogeneous quadratic polynomial in $x_1, x_2, \cdots, x_n$, which can be written
$$f = XAX^t = \sum_{i=1}^{n} \sum_{j=1}^{n} a_{ij} x_i x_j.$$

Quadratic forms will be discussed in some detail in Chapter 8. We note here, however, that when $i \neq j$, since $x_i x_j = x_j x_i$ the total coefficient of $x_i x_j$ in f can, without loss of generality, be assigned half to the term $x_i x_j$ and half to the term $x_j x_i$. In other words, every quadratic form in $x_1, x_2, \cdots, x_n$ can be written in the form XAX^t where the matrix A has the property that $a_{ij} = a_{ji}$ for all values of i and j. Such a matrix is said to be *symmetric*, and we conclude that *every quadratic form can be written XAX^t where A is a uniquely determined symmetric matrix*.

Matrix products also provide a very compact notation for writing systems of linear equations. For example, the student may easily verify that the system of equations

$$a_{11} x_1 + a_{12} x_2 + \cdots + a_{1n} x_n = c_1,$$
$$a_{21} x_1 + a_{22} x_2 + \cdots + a_{2n} x_n = c_2,$$
$$\cdots \cdots \cdots \cdots \cdots$$
$$a_{m1} x_1 + a_{m2} x_2 + \cdots + a_{mn} x_n = c_m$$

is equivalent to the matrix equation

(6) $$AX^t = C^t,$$

where A is the coefficient matrix (a_{ij}), X is the vector $(x_1, x_2, \cdots, x_n)$, and C is the vector $(c_1, c_2, \cdots, c_m)$. The problem of solving the equations becomes the problem of finding a vector X that will satisfy (6) when A and C are given. If $m = n$ and A is nonsingular, this can be done by multiplying each side of (6) on the left by A^{-1} and the solution may be written in the form $X^t = A^{-1}C^t$.

Finally, matrix methods are used extensively in the theory of linear differential equations. Suppose that $y_1, y_2, \cdots, y_n$ are n differentiable functions of a real variable x defined on a suitable interval, usually $0 \leq x < \infty$. The vector $Y = (y_1, y_2, \cdots, y_n)$ is then called a vector function of x, and its derivative is defined by

$$\frac{dY}{dx} = (y_1', y_2', \cdots, y_n'),$$

where $y_i' = \dfrac{dy_i}{dx}$. Every system of n linear first-order differential equations in $y_1, \cdots, y_n$ has the form

$$\frac{dy_1}{dx} = a_{11}y_1 + a_{12}y_2 + \cdots + a_{1n}y_n + c_1,$$

$$\cdots \cdots \cdots \cdots \cdots$$

$$\frac{dy_n}{dx} = a_{n1}y_1 + a_{n2}y_2 + \cdots + a_{nn}y_n + c_n,$$

where each coefficient a_{ij} is a function of x, and $C = (c_1, c_2, \cdots, c_n)$ is a vector function of x. If A is the matrix (a_{ij}), these equations can be written in the compact form

(7) $$\frac{dY^t}{dx} = AY^t + C^t.$$

Here the elements of the matrix A, like the coordinates of Y and C, are in general functions of x. There is no difficulty, however, in carrying over the rules of matrix algebra to this situation. The solution of systems of the form (7) will be discussed in Chapter 7.

Exercise 3.3

1. If

$$A = \begin{pmatrix} a & b \\ -b & a \end{pmatrix} \text{ and } B = \begin{pmatrix} c & d \\ -d & c \end{pmatrix}.$$

where a, b, c, d are real numbers, find $A \pm B$ and AB. Show that $AB = BA$. Assuming $B \neq 0$, find B^{-1} and AB^{-1}. Do these results suggest anything in your previous mathematical experience?

2. If $A = (a_{ij})$ is a nonsingular matrix of order n, the linear equations

$$\sum_{j=1}^{n} a_{ij} x_j = c_i \qquad (i = 1, 2, \cdots, n)$$

may be written as a matrix equation $AX^t = C^t$, whose solution is $X^t = A^{-1}C^t$. Show that this solution is a matrix formulation of Cramer's rule described in Section 16.

3. Solve the following systems of equations by finding the inverse of the coefficient matrices as indicated in Problem 2:

(a) $2x - y = 4$,
$3x + 2y = 7$.

(b) $x + y - 2z = 1$,
$2x - 7z = 3$,
$x + y - z = 5$.

4. Write down the matrix of each of the following bilinear forms, and verify that these forms may be written in the matrix notation described in Section 20:

(a) $x_1 y_1 + x_2 y_2 - x_3 y_3$.

(b) $2x_1 y_1 - 6x_1 y_2 + 2y_2 x_1 - x_2 y_2 + x_3 y_2 - 4x_3 y_3$.

5. Write down the matrix of each of the following quadratic forms, and verify that these forms may be written in the matrix notation described in Section 20:

(a) $x^2 + 2y^2 - 6z^2$.

(b) $ax^2 + 2bxy + cz^2$.

(c) $2x_1^2 + 3x_2^2 - x_3^2 + 6x_1 x_2 - 3x_1 x_3 + x_2 x_3$.

6. If

$$A = \begin{pmatrix} 2 & -1 & 3 \\ -1 & 5 & 2 \\ 3 & 2 & -1 \end{pmatrix},$$

$X = (x_1, x_2, x_3)$, and $Y = (y_1, y_2, y_3)$, write out in full the quadratic form XAX^t and the bilinear for XAY^t.

7. If $Y = T$ is a solution of (7), prove that every solution of (7) has the form $Y = S + T$, where $Y = S$ is a solution of the homogeneous system

(8) $$\frac{dY^t}{dx} = AY^t.$$

8. Prove that the solution vectors of the homogenous system (8) constitute a vector space over the real field in the sense of the abstract definition of a vector space given in Appendix 1.

21. ELEMENTARY TRANSFORMATIONS AND DETERMINATION OF RANK

We define six types of operations on a matrix A which we shall call *elementary transformations*. The six types of elementary transformations are the following:

1a Interchange of two rows of A.
1b Interchange of two columns of A.
2a Addition of a scalar multiple of one row vector of A to another.
2b Addition of a scalar multiple of one column vector of A to another.
3a Multiplication of a row vector of A by a nonzero scalar.
3b Multiplication of a column vector of A by a nonzero scalar.

It is obvious from the definition of rank that the application of elementary transformations cannot change the rank of a matrix. We state this fact formally as follows:

THEOREM 3.7. *The rank of a matrix is invariant under elementary transformations.*

DEFINITION. A matrix A is said to be *equivalent* to a matrix B if A can be transformed into B by a finite number of elementary transformations. We write $A \sim B$ for A is equivalent to B.

THEOREM 3.8. (a) If A is any matrix, $A \sim A$.
(b) If $A \sim B$, then $B \sim A$.
(c) If $A \sim B$ and $B \sim C$, then $A \sim C$.

Proof. Parts (a) and (c) are both obvious from the definition of equivalence. To prove (b) it is only necessary to note that each elementary transformation T can be "undone" by another elementary transformation T^{-1}, which will be called its *inverse*. For example, the inverse of the interchange of two rows is the interchange of the same two rows again. If T is the transformation that adds k times the jth row vector to the ith, then the inverse T^{-1} of T consists in adding $-k$ times the jth row vector to the ith. Finally, if T is the transformation that multiplies a row by a nonzero scalar k, then T^{-1} is the transformation that multiplies the same row by $1/k$. Now, if A is transformed into B by a succession of elementary transformations $T_1, T_2, \cdots, T_m$, the elementary transformations $T_m^{-1}, T_{m-1}^{-1}, \cdots, T_1^{-1}$ will clearly transform B into A, and therefore (b) follows.

The three laws (a), (b), and (c) of Theorem 3.8 are called, respectively, the reflexive, symmetric, and transitive laws. Any relation, defined among a set of numbers or other mathematical entities, that is reflexive,

Algebra of Matrices

symmetric, and transitive is called an *equivalence relation*. The simplest equivalence relation is, of course, *equality* but there are many others. Equivalence of matrices, is, by Theorem 3.8, an equivalence relation, and we shall meet others later. Still others are given in Exercise 3.4, Problems 6, 7, and 9.

We shall now describe a systematic procedure by which any matrix can be reduced by means of elementary transformations to a particularly simple form known as its *canonical form*. Since the rank of the canonical form will be at once apparent, and must, by Theorem 3.7, be equal to the rank of A, the procedure may be used to find the rank of any given matrix. We state the results as follows:

THEOREM 3.9. Every matrix A of rank r is equivalent to a matrix of the form

$$B = \begin{pmatrix} 1 & 0 & 0 & \cdots & 0 \\ 0 & 1 & 0 & \cdots & 0 \\ \cdot & \cdot & \cdot & \cdot & \cdot \\ 0 & \cdots & 1 & \cdots & 0 \\ 0 & \cdots & 0 & \cdots & 0 \\ \cdot & \cdot & \cdot & \cdot & \cdot \\ 0 & \cdot & \cdot & \cdot & 0 \end{pmatrix},$$

in which the first r elements in the main diagonal are equal to one and all other elements are equal to zero.

Proof. If A is a zero matrix so is B. If $A \neq O$, it contains a nonzero element that by application of elementary transformations of types $1a$ and $1b$ can be brought into the upper left-hand corner. Now, dividing the first row by the nonzero element in the upper left-hand corner (a transformation of type $3a$), we get a one in that place. By adding suitable scalar multiples of the first column vector to all the others we can now replace all elements of the first row, except the first, by zeros. Similarly by adding suitable multiples of the first row vector to all the others we reduce all elements of the first column, except the first, to zeros. All this is done by elementary transformations of type $2a$ and $2b$. We now have

$$A \sim \begin{pmatrix} 1 & 0 & 0 & \cdots & 0 \\ 0 & c_{22} & c_{23} & \cdots & c_{2n} \\ \cdot & \cdot & \cdot & \cdot & \cdot \\ 0 & c_{m2} & c_{m3} & \cdots & c_{mn} \end{pmatrix}.$$

Now, if not all the elements c_{ij} are zero, we can, by elementary transformations of type 1a and 1b, bring a nonzero element into the position of c_{22} without changing the first row and first column. Multiplying the second row vector then by $1/c_{22}$, we get a one in this position and again by elementary transformations of type 2a and 2b the elements $c_{23}, \cdots, c_{2n}$ and $c_{32}, \cdots, c_{m2}$ can be replaced by zeros. Continuing this process, we finally find that A is equivalent to a matrix of the form B. Since the rank of B is clearly equal to the number of ones in the main diagonal of B, it follows by Theorem 3.7 that this number is also the rank of A.

The matrix B is called the *canonical form* of A, and we have the rule that *the rank of any matrix is equal to the number of ones in the main diagonal of its canonical form*.

Since the canonical form of a matrix can actually be found by the method described in the proof of Theorem 3.9, this result gives a systematic procedure for finding the rank of a given matrix.

THEOREM 3.10. A necessary and sufficient condition that two matrices of the same type be equivalent is that they have the same rank, and hence the same canonical form.

Proof. If $A \sim B$, then A and B have the same rank by Theorem 3.7. Conversely, if A and B have the same rank and A' and B' are the canonical forms of A and B, we have, by Theorem 3.7, rank A' = rank A = rank B = rank B'. But from the form of A' and B' their ranks cannot be equal unless $A' = B'$. Hence by Theorem 3.8, since $A \sim A' = B' \sim B$, we have $A \sim B$ as required.

Exercise 3.4

1. Find the rank of each of the following matrices by reducing to canonical form:

(a) $\begin{pmatrix} 2 & 1 & 7 & 3 \\ 1 & 4 & 2 & 1 \\ 3 & 5 & 9 & 2 \end{pmatrix}$, (b) $\begin{pmatrix} 1 & 4 & 5 \\ 2 & 1 & 7 \\ 1 & -10 & -1 \end{pmatrix}$, (c) $\begin{pmatrix} 2 & -1 & 4 & 1 \\ 3 & 2 & 5 & -1 \\ 1 & 3 & 1 & -2 \\ 7 & 7 & 11 & -4 \end{pmatrix}$.

2. If $\mathscr{S}$ is the space generated by the vectors $(1, 5, 2, -2)$, $(3, 4, 1, 2)$, $(-3, 7, 4, -10)$ and $\mathscr{T}$ is the space generated by the vectors $(-2, 1, 1, -4)$ and $(2, 0, 5, -1)$, find the dimensions of the spaces $\mathscr{S}, \mathscr{T}, \mathscr{S} + \mathscr{T}$ and $\mathscr{S} \cap \mathscr{T}$. Find basis vectors for each of these spaces.

3. The rank of a system of homogeneous linear equations is defined to be the rank of the matrix of their coefficients. Find the rank of the system

$$x - 3y + w = 0,$$
$$2x + y + 3z - w = 0,$$
$$3x - 17y - 6z + 7w = 0,$$

and a set of basis vectors for the solution space.

4. Find the rank of the matrix

$$\begin{pmatrix} 1 & 3 & 2 & 0 & 1 \\ 9 & 2 & -1 & 6 & 4 \\ 7 & -4 & -5 & 6 & 2 \\ 17 & 1 & -4 & 12 & 7 \end{pmatrix}.$$

5. If A is the matrix of the coefficients of three nonhomogeneous equations in x, y, and z and if B is the augmented matrix of the system, list all the possible combinations r, r', where $r = $ rank A and $r' = $ rank B. Explain the geometric significance of each combination of ranks.

6. If m is a fixed nonzero integer and a and b are two integers, we define the relation $a \equiv b$ (mod m), (read a is congruent to b modulo m) to mean that $a - b$ is divisible by m. Prove that congruence modulo m is an equivalence relation.

7. If two real numbers x and y are defined to be equivalent whenever $x - y$ is a rational number, prove that this equivalence is an equivalence relation.

8. Think of other equivalence relations defined among the rational, real, or complex numbers.

9. A matrix A is said to be congruent to a matrix B if there exists a nonsingular matrix P such that $PAP^t = B$. Prove that congruence of matrices is an equivalence relation.

22. ELEMENTARY TRANSFORMATIONS AND MULTIPLICATION OF DETERMINANTS

THEOREM 3.11. *Every elementary transformation of a matrix A can be achieved by multiplying A by a nonsingular matrix.*

Proof. In proof of this theorem we shall merely give the nonsingular matrices which, when used as multipliers, will produce each of the six types of elementary transformations in A. The student should verify that the effect of each multiplication is as stated.

Suppose that A is of type (m, n). Let E_r be the matrix obtained from the unit matrix I_r by interchanging the ith and jth rows. Then $E_m A$ is the matrix obtained from A by interchanging the ith and jth rows, and AE_n is the matrix obtained from A by interchanging the ith and jth columns. The matrices E_m and E_n are nonsingular. Hence every elementary transformation of type 1a can be achieved by multiplying A on the left by a nonsingular matrix, and every elementary transformation of type 1b can be achieved by multiplying A on the right by a nonsingular matrix.

Now let F_r be the square matrix of order r which has all its elements in the main diagonal equal to one, the element in the ith row and jth ($i \neq j$) column equal to k, and all other elements equal to zero. The matrix $F_m A$ is then the matrix obtained from A by adding k times the jth row of A to the ith row, and AF_n is the matrix obtained from A by adding k times the ith column of A to the jth column. The matrices F_m and F_n are nonsingular, and therefore any elementary transformation of type 2a can be achieved by multiplying A on the left by a nonsingular matrix, and any elementary transformation of type 2b can be achieved by multiplying A on the right by a nonsingular matrix.

Finally let G_r be a square matrix of order r in which all the main diagonal elements are equal to one except the one in the ith row and ith column which is equal to k, and in which all the other elements are equal to zero. Then $G_m A$ is the matrix obtained from A by multiplying the ith row by k, and AG_n is the matrix obtained from A by multiplying the ith column by k. If $k \neq 0$, G_m and G_n are nonsingular, and hence any elementary transformation of type 3a can be achieved by multiplying A on the left by a nonsingular matrix, and any elementary transformation of type 3b can be achieved by multiplying A on the right by a nonsingular matrix.

The proof of Theorem 3.11 is now complete but we have actually proved more than was stated. Since the product of two nonsingular matrices is nonsingular, the above proof also yields the following:

THEOREM 3.12. *If A is an arbitrary matrix there exist nonsingular matrices P and Q such that $PAQ = A'$, where A' is the canonical form of A.*

We note that the matrices P and Q in Theorem 3.12 are by no means uniquely determined. For example, if R is any nonsingular matrix that commutes with A' (i.e., $A'R = RA'$), then $(R^{-1}P)A(QR) = R^{-1}A'R = A'$. If A is nonsingular, $A' = I$, and in this case $(R^{-1}P)A(QR) = A'$ for any nonsingular matrix R.

If by an *elementary matrix* we understand any matrix of the type E, F, or G described in the proof of Theorem 3.11, we can deduce the following consequences of Theorem 3.12.

Algebra of Matrices 75

COROLLARY 1. Every nonsingular matrix is a product of elementary matrices.

Proof. If A is nonsingular its canonical form is the unit matrix I. By Theorem 3.8(b), I can be transformed into A by elementary transformations, and hence, by Theorem 3.12, $A = PIQ = PQ$, where both P and Q are products of elementary matrices.

COROLLARY 2. A necessary and sufficient condition that A be equivalent to B is that there exist nonsingular matrices M and N such that $MAN = B$.

Proof. If $A \sim B$, both have the same canonical form, and hence there exist nonsingular matrices P, Q, R, and S such that $PAQ = RBS$. Hence $B = R^{-1}PAQS^{-1} = MAN$, where both $M = R^{-1}P$ and $N = QS^{-1}$ are nonsingular by the corollary to Theorem 3.4. Conversely, if $B = MAN$, where M and N are nonsingular, it follows by Corollary 1 that both M and N are products of elementary matrices. Hence A can be transformed into B by a sequence of elementary transformations, and $A \sim B$.

Our next objective is to prove the multiplication theorem for determinants, namely, that, if A and B are any two square matrices of order n, then $|AB| = |A| \, |B|$. We shall prove it first for a special case and then deduce the general theorem.

LEMMA. If B is any elementary matrix of order n and A an arbitrary matrix of order n, then

$$|AB| = |A| \, |B| \quad \text{and} \quad |BA| = |B| \, |A|.$$

Proof. If B is an elementary matrix of type E, then $|B| = -1$ and $|AB| = |BA| = -|A|$, since multiplication by B simply interchanges two rows or two columns of A. Hence the lemma holds in this case. If B is of type F, then $|B| = 1$, and multiplication by B adds a scalar multiple of one row or column of A to another row or column. Hence $|AB| = |BA| = |A| = |A| \, |B|$. Finally, if B is of type G, then $|B| = k$, and multiplication by B multiplies one row or one column of A by a constant k. Hence $|AB| = |BA| = k|A| = |B| \, |A|$. The proof of the lemma is therefore complete.

THEOREM 3.13. If A and B are any square matrices of order n, then $|AB| = |A| \, |B|$.

Proof. We note first that, if either A or B is singular, then AB is singular by Theorem 3.2, and $|AB| = |A| \, |B| = 0$ by Theorem 2.2. The theorem therefore holds in this case. If both A and B are nonsingular, then, by Corollary 1, Theorem 3.12, $A = C_1 C_2 \cdots C_r$, where C_1, C_2,

$\cdots$, C_r are elementary matrices. Hence by successive application of the preceding lemma

$$\begin{aligned}|AB| &= |C_1 C_2 \cdots C_r B| \\ &= |C_1|\ |C_2|\ \cdots\ |C_r|\ |B| \\ &= |C_1 C_2 \cdots C_r|\ |B| \\ &= |A|\ |B|.\end{aligned}$$

The theorem is therefore proved.

Exercise 3.5

1. If
$$A = \begin{pmatrix} 1 & 0 & 2 \\ 0 & 3 & -1 \\ 2 & 3 & 3 \end{pmatrix},$$
find nonsingular matrices P and Q such that PAQ is in canonical form. What is the rank of A?

2. If A is a nonsingular matrix and $|A| = d$, prove that $|A^{-1}| = \dfrac{1}{d}$.

3. If $A = (a_{ij})$ is a square matrix of order n and $B = (|A_{ji}|^*)$, prove that $|B| = |A|^{n-1}$. (*Hint:* see Problem 2, Exercise 3.2.)

4. If $A = (a_{ij})$ is a symmetric matrix of order n (i.e., $a_{ij} = a_{ji}$ for all values of i and j), prove that there exists a nonsingular matrix P such that PAP^t is the canonical form of A. (*Hint:* show that the reduction to canonical form can be performed in such a way that an elementary transformation on rows is always followed by the same elementary transformation on columns, and that these can be achieved by multiplying A on the left by an elementary matrix E and on the right by E^t.)

23. MULTIPLICATION OF PARTITIONED MATRICES

It is frequently convenient to partition a matrix into submatrices and to consider it as a matrix whose elements are themselves these submatrices. For example, the matrix

$$A = \begin{pmatrix} a_{11} & a_{12} & a_{13} & a_{14} & a_{15} \\ a_{21} & a_{22} & a_{23} & a_{24} & a_{25} \\ \hline a_{31} & a_{32} & a_{33} & a_{34} & a_{35} \\ \hline a_{41} & a_{42} & a_{43} & a_{44} & a_{45} \end{pmatrix}$$

Algebra of Matrices

may be partitioned as shown and written in the form

$$\begin{pmatrix} A_{11} & A_{12} & A_{13} \\ A_{21} & A_{22} & A_{23} \end{pmatrix},$$

where

$$A_{11} = \begin{pmatrix} a_{11} & a_{12} \\ a_{21} & a_{22} \end{pmatrix}, \quad A_{12} = \begin{pmatrix} a_{13} & a_{14} \\ a_{23} & a_{24} \end{pmatrix}, \quad A_{13} = \begin{pmatrix} a_{15} \\ a_{25} \end{pmatrix}$$

$$A_{21} = \begin{pmatrix} a_{31} & a_{32} \\ a_{41} & a_{42} \end{pmatrix}, \quad A_{22} = \begin{pmatrix} a_{33} & a_{32} \\ a_{43} & a_{44} \end{pmatrix}, \quad A_{23} = \begin{pmatrix} a_{35} \\ a_{45} \end{pmatrix}.$$

Now, if B is any matrix with five rows, the product AB is defined. To be definite, suppose

$$B = \begin{pmatrix} b_{11} & b_{12} & b_{13} \\ b_{21} & b_{22} & b_{23} \\ \hline b_{31} & b_{32} & b_{33} \\ b_{41} & b_{42} & b_{43} \\ \hline b_{51} & b_{52} & b_{53} \end{pmatrix} = \begin{pmatrix} B_{11} & B_{12} \\ B_{21} & B_{22} \\ B_{31} & B_{32} \end{pmatrix}.$$

Now let B be partitioned in the manner shown so that its row partitions are spaced in the same way as the column partitions of A. Then we can multiply the two matrices in *partitioned form*. That is to say, we can write

$$AB = \begin{pmatrix} A_{11}B_{11} + A_{12}B_{21} + A_{13}B_{31} & A_{11}B_{12} + A_{12}B_{22} + A_{13}B_{32} \\ A_{21}B_{11} + A_{22}B_{21} + A_{23}B_{31} & A_{21}B_{12} + A_{22}B_{22} + A_{23}B_{32} \end{pmatrix}$$

as a partitioned form of the product AB. The student should verify, by writing out the products in full, first that all the matrix products and sums used in this multiplication are actually defined, and second that the above product of the partitioned forms of A and B does actually give the product AB.

This rule for multiplying partitioned matrices is general and may be stated as follows: If

$$A = \begin{pmatrix} A_{11} & A_{12} & \cdots & A_{1n} \\ \cdot & \cdot & & \cdot \\ A_{m1} & A_{m2} & \cdots & A_{mn} \end{pmatrix} \quad \text{and} \quad B = \begin{pmatrix} B_{11} & B_{12} & \cdots & B_{1h} \\ \cdot & \cdot & & \cdot \\ B_{n1} & B_{n2} & \cdots & B_{nh} \end{pmatrix},$$

where A_{ij} is a matrix of type (r_i, s_j) and B_{jk} is a matrix of type (s_j, t_k), then

$$AB = \left(\sum_{j=1}^{n} A_{ij}B_{jk}\right) \qquad (i = 1, \cdots, m, k = 1, \cdots, h).$$

A formal proof of this rule will not be given but the student should test its truth by working several examples of the type indicated above.

CHAPTER FOUR

Further Geometry of Real Vector Spaces

24. ORTHOGONAL COMPLEMENTS AND ORTHOGONAL PROJECTIONS

Let $\mathscr{S}$ be any subspace of $\mathscr{V}_n(\mathscr{R})$. A vector X is said to be orthogonal to $\mathscr{S}$ if it is orthogonal to every vector of $\mathscr{S}$. It is easy to see that the vectors that are orthogonal to $\mathscr{S}$ constitute a vector space, for, if X and Y are orthogonal to every vector of $\mathscr{S}$, so also is every linear combination $aX + bY$.

DEFINITION. The space $\mathscr{T}$ consisting of all vectors of $\mathscr{V}_n(\mathscr{R})$ that are orthogonal to $\mathscr{S}$ is called the *orthogonal complement* of $\mathscr{S}$ in $\mathscr{V}_n(\mathscr{R})$.

THEOREM 4.1. If $\mathscr{S}$ is a subspace of $\mathscr{V}_n(\mathscr{R})$ of dimension r, and if $\mathscr{T}$ is the orthogonal complement of $\mathscr{S}$ in $\mathscr{V}_n(\mathscr{R})$, then $\mathscr{S} \cap \mathscr{T} = 0$, $\mathscr{T}$ has dimension $n - r$, and $\mathscr{S} + \mathscr{T} = \mathscr{V}_n(\mathscr{R})$.

Proof. Since every vector of $\mathscr{T}$ is orthogonal to every vector of $\mathscr{S}$ it follows that $\mathscr{S} \cap \mathscr{T}$ consists of the zero vector only, since this is the only vector that is orthogonal to itself.

By Theorem 1.17 there exists a normal orthogonal basis $E_1, E_2, \cdots, E_n$ of $\mathscr{V}_n(\mathscr{R})$ such that $E_1, E_2, \cdots, E_r$ is a basis of $\mathscr{S}$. The vectors $E_{r+1}, \cdots, E_n$ generate a space $\mathscr{T}'$, of dimension $n - r$, which is certainly contained in $\mathscr{T}$. Hence, if $\mathscr{T}$ has dimension t, we have $n - r \le t$. However, since $\mathscr{S} \cap \mathscr{T} = O$, it follows from Theorem 1.10 that $t \le n - r$. Hence $t = n - r$ and $\mathscr{T}' = \mathscr{T}$ by Problem 3, Exercise 1.8. Again by Theorem 1.10, $\mathscr{S} + \mathscr{T}$ has dimension n, and hence $\mathscr{S} + \mathscr{T} = \mathscr{V}_n(\mathscr{R})$.

THEOREM 4.2. If $\mathscr{T}$ is the orthogonal complement of $\mathscr{S}$ in $\mathscr{V}_n(\mathscr{R})$, then every vector X of $\mathscr{V}_n(\mathscr{R})$ can be written in one and only one way in the form $X = S + T$, where $S \in \mathscr{S}$ and $T \in \mathscr{T}$.

Linear Algebra for Undergraduates

Proof. Since $\mathscr{V}_n(\mathscr{R}) = \mathscr{S} + \mathscr{T}$, every vector X in $\mathscr{V}_n(\mathscr{R})$ can be written in the required form. If $X = S + T = S_1 + T_1$ are two such representations, then $S - S_1 = T_1 - T$. Since $S - S_1 \in \mathscr{S}$ and $T_1 - T \in \mathscr{T}$, both must belong to $\mathscr{S} \cap \mathscr{T}$. Hence $S - S_1 = T_1 - T = O$ and $S_1 = S$ and $T_1 = T$. The representation $X = S + T$ is therefore unique.

It should perhaps be noted that the proof of Theorem 4.2 makes no use of the fact that $\mathscr{S}$ and $\mathscr{T}$ are orthogonal. The same proof shows that, if $\mathscr{S}$ and $\mathscr{T}$ are any two spaces for which $\mathscr{S} \cap \mathscr{T} = O$, then every vector of $\mathscr{S} + \mathscr{T}$ has a unique representation in the form $S + T$ where $S \in \mathscr{S}$ and $T \in \mathscr{T}$. The theorem has been stated for orthogonal spaces whose sum is $\mathscr{V}_n(\mathscr{R})$ because this is the form in which we shall make use of the result.

THEOREM 4.3. *If $\mathscr{T}$ is the orthogonal complement of $\mathscr{S}$ in $\mathscr{V}_n(\mathscr{R})$, then $\mathscr{S}$ is the orthogonal complement of $\mathscr{T}$ in $\mathscr{V}_n(\mathscr{R})$.*

Proof. Since every vector of $\mathscr{S}$ is certainly orthogonal to $\mathscr{T}$, it is only necessary to prove that every vector of $\mathscr{V}_n(\mathscr{R})$ that is orthogonal to $\mathscr{T}$ belongs to $\mathscr{S}$. Suppose that X is orthogonal to $\mathscr{T}$. By Theorem 4.2 $X = S + T$, where $S \in \mathscr{S}$ and $T \in \mathscr{T}$. Hence

$$X \cdot T = S \cdot T + T \cdot T.$$

But $X \cdot T = 0$ and $S \cdot T = 0$, whence $T \cdot T = 0$ and $T = O$. Hence $X = S$, and therefore $X \in \mathscr{S}$ as required.

DEFINITION. *If T is the orthogonal complement of $\mathscr{S}$ in $\mathscr{V}_n(\mathscr{R})$ and if $X = S + T$, where $S \in \mathscr{S}$ and $T \in \mathscr{T}$, we shall call S the orthogonal projection (or simply the projection) of the vector X onto the space $\mathscr{S}$. Similarly T is the projection of X on the space $\mathscr{T}$.* By Theorem 4.2 these projections are uniquely determined.

The reason for the above definition will become apparent if we consider an example from three-dimensional space. Let $\mathscr{S}$ be a one-dimensional subspace of $\mathscr{V}_3(\mathscr{R})$, that is, the set of all vectors lying in a fixed line l through the origin O. The orthogonal complement $\mathscr{T}$ of $\mathscr{S}$ is therefore the space of all vectors orthogonal to $\mathscr{S}$ or all vectors lying in the plane p which is perpendicular to l and passes through O. Now let X be an arbitrary vector in $\mathscr{V}_3(\mathscr{R})$ with geometric representation OP. To write X in the form $S + T$, where $S \in \mathscr{S}$ and $T \in \mathscr{T}$, we must construct a parallelogram of which OP is a diagonal and whose sides OQ and OR lie in the line l and in the plane p, respectively. Since S and T are orthogonal it is clear that the parallelogram is a rectangle. Moreover its sides are the orthogonal projections, in the ordinary geometric sense, of the line

OP onto the line l and the plane p. These two projections are the geometric representations of the vectors S and T for which $X = S + T$.

25. EQUATIONS OF A SUBSPACE OF $\mathscr{V}_n(\mathscr{R})$

In Theorem 1.2 it was shown that every system of linear homogeneous equations in n unknowns with real coefficients defines a subspace of $\mathscr{V}_n(\mathscr{R})$ which we called the solution space of the equations. This space consists of all vectors whose coordinates satisfy the given system of equations. From Theorem 2.5 we know that if the number of *independent* equations in the system is r, then the dimension of the solution space is $n - r$. In this section we shall consider the converse problem, namely, given a subspace of $\mathscr{V}_n(\mathscr{R})$, is it the solution space of a system of homogeneous equations and, if so, how can such a system of equations be found? The first question is answered in the affirmative by the following theorem and the proof of this theorem will provide an answer to the second.

THEOREM 4.4. Every r-dimensional subspace $\mathscr{S}$ of $\mathscr{V}_n(\mathscr{R})$ is the solution space of a set of $n - r$ independent linear homogeneous equations in $x_1, x_2, \cdots, x_n$.

Proof. Let $\mathscr{T}$ be the orthogonal complement of $\mathscr{S}$ in $\mathscr{V}_n(\mathscr{R})$. By Theorem 4.1, $\mathscr{T}$ has dimension $n - r$. Let

$$T_1 = (b_{11}, b_{12}, \cdots, b_{1n}),$$
$$\cdots \cdots \cdots \cdots$$
$$T_{n-r} = (b_{n-r1}, b_{n-r2}, \cdots, b_{n-rn}),$$

be a basis of $\mathscr{T}$. Since, by Theorem 4.3, $\mathscr{S}$ is the orthogonal complement of $\mathscr{T}$, it follows that $\mathscr{S}$ is the solution space of the equations

(1)
$$b_{11}x_1 + b_{12}x_2 + \cdots + b_{1n}x_n = 0,$$
$$\cdots \cdots \cdots \cdots$$
$$b_{n-r1}x_1 + b_{n-r2}x_2 + \cdots + b_{n-rn}x_n = 0.$$

These $n - r$ equations are independent since the vectors $T_1, T_2, \cdots, T_{n-r}$ are a basis of $\mathscr{T}$ and are therefore linearly independent. This completes the proof.

Equations 1 will be called equations of the space $\mathscr{S}$. They are not uniquely determined since there is a set of equations of $\mathscr{S}$ corresponding to every basis $T_1, T_2, \cdots, T_{n-r}$ of $\mathscr{T}$. The relation of a subspace and its equations is exactly that of elementary analytic geometry. The coordinates of every vector of $\mathscr{S}$ satisfy each of equations 1, and conversely every vector whose coordinates satisfy equations 1 belongs to $\mathscr{S}$. We now have two ways of defining a space $\mathscr{S}$. It may be defined by its equations or by a set of basis vectors. If equations of a space are given, a set of basis vectors may be found as in the proof of Theorem 2.5. On the other hand if the basis vectors of $\mathscr{S}$ are given, equations of the orthogonal complement $\mathscr{T}$ of $\mathscr{S}$ can immediately be written down. From these a set of basis vectors of $\mathscr{T}$ can be found as in Theorem 2.5, and from them equations of $\mathscr{S}$ are found.

A set of basis vectors of a space $\mathscr{S}$ also enables us to define $\mathscr{S}$ by a set of *parametric equations*. Suppose, for example, that $\mathscr{S}$ has dimension r and that the vectors

$$S_1 = (a_{11}, a_{12}, \cdots, a_{1n}),$$
$$\cdots \cdots \cdots \cdots$$
$$S_r = (a_{r1}, a_{r2}, \cdots, a_{rn})$$

constitute a basis of $\mathscr{S}$. The general vector X of $\mathscr{S}$ can then be written in the form

(2) $$X = p_1 S_1 + p_2 S_2 + \cdots + p_r S_r.$$

This is the vector form of the parametric equations of $\mathscr{S}$. As the parameters $p_1, p_2, \cdots, p_r$ vary independently over the real numbers, X varies over the space $\mathscr{S}$. If $X = (x_1, x_2, \cdots, x_n)$ and we substitute for $S_1, S_2, \cdots, S_r$, equation 2 yields, on equating corresponding coordinates,

$$x_1 = a_{11} p_1 + a_{21} p_2 + \cdots + a_{r1} p_r,$$
$$x_2 = a_{12} p_1 + a_{22} p_2 + \cdots + a_{r2} p_r,$$
$$\cdots \cdots \cdots \cdots \cdots$$
$$x_n = a_{1n} p_1 + a_{2n} p_2 + \cdots + a_{rn} p_r.$$

These are the parametric equations of $\mathscr{S}$ in the usual sense: substitution of any set of real values for the parameters $p_1, p_2, \cdots, p_r$ gives a vector $(x_1, x_2, \cdots, x_n)$ which belongs to $\mathscr{S}$, and conversely every vector of $\mathscr{S}$ is obtained in this way.

Exercise 4.1

1. Specialize the result of Theorem 4.4 to one- and two-dimensional subspaces of $\mathscr{V}_3(\mathscr{R})$, and thus obtain familiar theorems of three-dimensional analytic geometry.

Geometry of Vector Spaces   83

2. Find basis vectors and parametric equations for the solution space of each of the following systems of equations:

(a) $x - 2y + 3z = 0.$ (b) $x + y - z = 0,$ (c) $x - y + 2z + w = 0,$
$\qquad\qquad\qquad\qquad\quad 2x - y + 2z = 0.\qquad 2x - y + z - 2w = 0.$

3. Find equations of the subspace $\mathscr{S}$ of $\mathscr{V}_4(\mathscr{R})$ which is generated by the vectors $(2, 1, 3, -1)$, $(5, 2, 1, 1)$, and $(-2, 3, 1, 4)$.

4. An $(n-1)$-dimensional subspace of $\mathscr{V}_n(\mathscr{R})$ is called a *hyperplane*. Prove that every r-dimensional subspace of $\mathscr{V}_n(\mathscr{R})$ is the intersection of $n - r$ hyperplanes.

5. Find a normal orthogonal basis for the solution space of the equations:

$$x + 2y + 2z - w = 0,$$
$$2x + 3y - z + 4w = 0.$$

6. Find the orthogonal projection of the vector $(2, 1, -3)$ onto the plane $2x - 3y + 5z = 0$.

7. Find the orthogonal projection of the vector $(2, -1, 2)$ onto the space generated by the vector $(1, -1, 3)$.

8. Find a basis for the solution space of the equations

$$x + 2y - z - w = 0,$$
$$3x - y + z + 4w = 0$$

and also for the orthogonal complement of this space.

9. Find the projections of the vector $(1, 4, 2, 2)$ onto the two orthogonally complementary spaces of Problem 8.

26. VOLUME IN n-DIMENSIONAL SPACE

The concept of length of a vector has already been generalized to spaces of any number of dimensions. We now seek similar generalizations of the ideas of area and volume. We start by deriving a simple formula for the area of a parallelogram. Both the formula and the method of derivation will suggest immediate generalization to n-dimensional space.

Let $X = (x_1, x_2)$ and $Y = (y_1, y_2)$ be any two linearly independent vectors in the plane, and consider the parallelogram of which X and Y are two edges. By elementary geometry the area of this parallelogram is the product of the base and the altitude. If we choose the vector X as base, the altitude is equal to the length of the projection Z of Y onto the line through O perpendicular to X. But Z is then the projection of Y onto the orthogonal complement in $\mathscr{V}_2(\mathscr{R})$ of the space $\{X\}$ generated by X. Hence, by Theorem 4.2, we may write

$$Y = kX + Z,$$

where $X \cdot Z = 0$ and k and Z are uniquely determined. Forming the inner products $Y \cdot X$ and $Y \cdot Z$, we get, since $X \cdot Z = 0$,

$$X \cdot Y = k\|X\|^2,$$

whence

(3) $$k = \frac{X \cdot Y}{\|X\|^2},$$

and

(4) $$Y \cdot Z = kX \cdot Z + Z \cdot Z = \|Z\|^2.$$

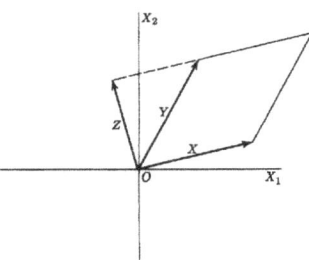

Figure 6

Now, if a is the area of the parallelogram, we have, using equations 3 and 4,

$$a^2 = \|X\|^2\|Z\|^2 = \|X\|^2(Y \cdot Z) = \|X\|^2 Y \cdot (Y - kX)$$
$$= \|X\|^2\|Y\|^2 - (X \cdot Y)^2.$$

We now have the area a in terms of the vectors X and Y. Our formula can also be written in the form

(5) $$a^2 = \begin{vmatrix} X \cdot X & X \cdot Y \\ Y \cdot X & Y \cdot Y \end{vmatrix}$$

or

$$a^2 = \begin{vmatrix} x_1 & x_2 \\ y_1 & y_2 \end{vmatrix} \begin{vmatrix} x_1 & y_1 \\ x_2 & y_2 \end{vmatrix} = \begin{vmatrix} x_1 & x_2 \\ y_1 & y_2 \end{vmatrix}^2.$$

Geometry of Vector Spaces

We can now state our results as

THEOREM 4.5. The area of the parallelogram whose edges are vectors $X = (x_1, x_2)$ and $Y = (y_1, y_2)$ is given either by the positive square root of the determinant

$$\begin{vmatrix} X \cdot X & X \cdot Y \\ Y \cdot X & Y \cdot Y \end{vmatrix}$$

or by the absolute value of the determinant

$$\begin{vmatrix} x_1 & x_2 \\ y_1 & y_2 \end{vmatrix}.$$

Exercise 4.2

1. Find the area of the parallelograms whose vertices are
   (a) (0, 0), (1, − 2), (3, 4), and (4, 2).
   (b) (2, 1), (− 1,3), (3, − 6), and (0, − 4).
   (c) (− 1, 3), (1, 5), (3, 2), and (5, 4).

2. Prove algebraically that the parallelogram whose edges are the vectors X and Y is equal in area to the parallelogram whose edges are X and $X + Y$. Interpret this result geometrically.

3. Using the method of Theorem 4.5, prove that equation 5 also gives the area of the parallelogram in space whose edges are the vectors $X = (x_1, x_2, x_3)$ and $Y = (y_1, y_2, y_3)$.

4. Find the area of the parallelograms whose vertices are
   (a) (0, 0, 0), (1, − 2, 2), (2, 3, 3), and (3, 1, 5).
   (b) (2, 2, 1), (3, 0, 6), (4, 1, 5), and (1, 1, 2).

Now suppose that $X = (x_1, x_2, x_3)$, $Y = (y_1, y_2, y_3)$, and $Z = (z_1, z_2, z_3)$ are any three linearly independent vectors in $\mathscr{V}_3(\mathscr{R})$. By passing through the terminal point of each of these vectors a plane parallel to the plane of the other two, we can complete a parallelepiped the lengths of whose edges are $\|X\|$, $\|Y\|$, and $\|Z\|$. The volume of this parallelepiped is the product of the area of a base, say the parallelogram defined by X and Y, and the altitude, which is the length of the projection of Z onto a line through the origin perpendicular to the plane of X and Y. Thus, the altitude is the length of the projection of Z onto the orthogonal complement in $\mathscr{V}_3(\mathscr{R})$ of the space $\{X, Y\}$ generated by X and Y.

To find this projection W of Z, we write

$$Z = W + bX + cY,$$

where, by Theorem 4.2, W, b, c are uniquely determined. Forming inner products of Z with X, Y, and W, respectively, we get, since $X\cdot W = Y\cdot W = 0$,

(6)
$$X\cdot Z = b\|X\|^2 + c(X\cdot Y),$$
$$Y\cdot Z = b(Y\cdot X) + c\|Y\|^2,$$
$$W\cdot Z = \|W\|^2.$$

By the result of Problem 3, Exercise 4.2, the area a of the base of the parallelepiped is given by (5), so that, on solving the first two equations of (6) for b and c, we get

$$a^2 b = \begin{vmatrix} X\cdot Z & X\cdot Y \\ Y\cdot Z & Y\cdot Y \end{vmatrix} = -\begin{vmatrix} X\cdot Y & X\cdot Z \\ Y\cdot Y & Y\cdot Z \end{vmatrix}$$

$$a^2 c = \begin{vmatrix} X\cdot X & X\cdot Z \\ Y\cdot X & Y\cdot Z \end{vmatrix}$$

If v is the required volume, we have, using the third equation of (6),

$$v^2 = a^2\|W\|^2 = a^2(W\cdot Z) = a^2 Z\cdot(Z - bX - cY) = a^2(Z\cdot Z - bX\cdot Z - cY\cdot Z).$$

Now, substituting for a^2, b, and c, we find

$$v^2 = X\cdot Z \begin{vmatrix} X\cdot Y & X\cdot Z \\ Y\cdot Y & Y\cdot Z \end{vmatrix} - Y\cdot Z \begin{vmatrix} X\cdot X & X\cdot Z \\ Y\cdot X & Y\cdot Z \end{vmatrix} + Z\cdot Z \begin{vmatrix} X\cdot X & X\cdot Y \\ Y\cdot X & Y\cdot Y \end{vmatrix}$$

$$= \begin{vmatrix} X\cdot X & Y\cdot X & Z\cdot X \\ X\cdot Y & Y\cdot Y & Z\cdot Y \\ X\cdot Z & Y\cdot Z & Z\cdot Z \end{vmatrix} = \begin{vmatrix} x_1 & x_2 & x_3 \\ y_1 & y_2 & y_3 \\ z_1 & z_2 & z_3 \end{vmatrix} \cdot \begin{vmatrix} x_1 & y_1 & z_1 \\ x_2 & y_2 & z_2 \\ x_3 & y_3 & z_3 \end{vmatrix} = \begin{vmatrix} x_1 & x_2 & x_3 \\ y_1 & y_2 & y_3 \\ z_1 & z_2 & z_3 \end{vmatrix}^2$$

Hence we have proved

THEOREM 4.6. *The volume of the parallelepiped defined by the three vectors* $X = (x_1, x_2, x_3)$, $Y = (y_1, y_2, y_3)$, *and* $Z = (z_1, z_2, z_3)$ *is equal to the positive square root of the determinant*

$$\begin{vmatrix} X\cdot X & Y\cdot X & Z\cdot X \\ X\cdot Y & Y\cdot Y & Z\cdot Y \\ X\cdot Z & Y\cdot Z & Z\cdot Z \end{vmatrix}$$

or to the absolute value of the determinant

$$\begin{vmatrix} x_1 & x_2 & x_3 \\ y_1 & y_2 & y_3 \\ z_1 & z_2 & z_3 \end{vmatrix}$$

Geometry of Vector Spaces

The results of Theorems 4.5 and 4.6 can be generalized to n-dimensional space. First we define what is meant by an r-dimensional parallelepiped in $\mathscr{V}_n(\mathscr{R})$. To this end we note that the parallelogram whose sides are the two vectors X and Y in $\mathscr{V}_2(\mathscr{R})$ "contains" all vectors of the form $aX + bY$, where $0 \leq a \leq 1$ and $0 \leq b \leq 1$. By this we mean that all these vectors have geometric representations whose terminal points lie inside or on the boundary of the parallelogram. In fact, a necessary and sufficient condition that a vector have its terminal point in or on the parallelogram is that it be of the above form. Similarly, if X, Y, Z are three linearly independent vectors in 3-space, the parallelepiped whose edges are the vectors X, Y, and Z can be characterized as that portion of space that contains all vectors of the form $aX + bY + cZ$, where $0 \leq a \leq 1$, $0 \leq b \leq 1$, and $0 \leq c \leq 1$. The vertices of this parallelepiped are the terminal points of the eight vectors $e_1 X + e_2 Y + e_3 Z$, where each e_i is either zero or one. It is now clear how to formulate the definition of an r-dimensional parallelepiped.

DEFINITION. Any r linearly independent vectors $X_1, X_2, \cdots, X_r$ of $\mathscr{V}_n(\mathscr{R})$ are said to define an *r-dimensional parallelepiped* which consists of all vectors of the form $a_1 X_1 + a_2 X_2 + \cdots + a_r X_r$, where $0 \leq a_i \leq 1$ ($i = 1, 2, \cdots, r$).

An r-dimensional parallelepiped has 2^r *vertices*, namely, the "terminal points" of the 2^r vectors $e_1 X_1 + e_2 X_2 + \cdots + e_r X_r$, where each e_i is zero or one. Two vertices are said to be adjacent if they are terminal points of two of the above vectors whose difference is X_i ($i = 1, 2, \cdots, r$). Any two adjacent vertices determine an edge of the parallelepiped. Thus the number of edges is $\dfrac{r \cdot 2^r}{2}$ or $r \cdot 2^{r-1}$ since r edges meet at each of the 2^r vertices but each edge contains exactly two vertices. The student should check this result for a parallelogram and for a three-dimensional parallelepiped. Any two edges that meet in one vertex define a two-dimensional face of the parallelepiped. The number of two-dimensional faces meeting at each vertex is therefore $\binom{r}{2}$ or $\dfrac{r(r-1)}{2}$. The total number of two-dimensional faces is therefore $\dfrac{1}{4}\binom{r}{2} 2^r = r(r-1) 2^{r-3}$ since each such face contains four vertices. By similar methods we could count the 3-, 4-, $\cdots$, $(r-1)$-dimensional faces of the parallelepiped. If the student has difficulty in "visualizing" these higher-dimensional figures, this should not worry him. He should remember that all our theorems for higher-dimensional space are essentially algebraic in nature. The geometric language is retained partly to preserve the relation between algebra and

88   Linear Algebra for Undergraduates

geometry, which is familiar in the three-dimensional case, and partly for the additional light it sheds on the algebraic relationships.

Our next task is to define the r-volume of an r-dimensional parallelepiped in such a way that in $\mathscr{V}_3(\mathscr{R})$ 1-volume will mean length, 2-volume, area, and 3-volume, volume in the ordinary sense. We give the definition by induction on r. If X is any nonzero vector in $\mathscr{V}_n(\mathscr{R})$, its 1-volume is defined to be its length, namely $\sqrt{X \cdot X}$. Now suppose that the $(r-1)$-volume of an $(r-1)$-dimensional parallelepiped has been defined. Let Π be the r-dimensional parallelepiped defined by the r linearly independent vectors $X_1, X_2, \cdots, X_r$. By the "base" of Π we shall understand the $(r-1)$-dimensional parallelepiped defined by $X_1, X_2, \cdots, X_{r-1}$, and by the "altitude" of Π we shall mean the length of the projection of X_r onto the orthogonal complement in $\mathscr{V}_n(\mathscr{R})$ of the space $\{X_1, X_2, \cdots, X_{r-1}\}$. By our induction assumption the $(r-1)$-volume of the base is already defined. We therefore define the r-volume of Π to be the product of the $(r-1)$-volume of its base and its altitude. We shall now use this definition to derive a formula for the r-volume of an r-dimensional parallelepiped which will include Theorems 4.5 and 4.6 as special cases.

THEOREM 4.7. *The r-volume of the r-dimensional parallelepiped Π defined by the r linearly independent vectors $X_1, X_2, \cdots, X_r$ of $\mathscr{V}_n(\mathscr{R})$ is the positive square root of the determinant*

(7)
$$\begin{vmatrix} X_1 \cdot X_1 & X_1 \cdot X_2 & \cdots & X_1 \cdot X_r \\ X_2 \cdot X_1 & X_2 \cdot X_2 & \cdots & X_2 \cdot X_r \\ \cdots & \cdots & \cdots & \cdots \\ X_r \cdot X_1 & X_r \cdot X_2 & \cdots & X_r \cdot X_r \end{vmatrix}$$

Proof. The theorem is true for $r = 1$ by the definition of length (or 1-volume) of a vector. We assume the truth of the theorem for an $(r-1)$-dimensional parallelepiped so that, if the $(r-1)$-volume of the base of Π is b, we have

$$b^2 = \begin{vmatrix} X_1 \cdot X_1 & X_1 \cdot X_2 & \cdots & X_1 \cdot X_{r-1} \\ X_2 \cdot X_1 & X_2 \cdot X_2 & \cdots & X_2 \cdot X_{r-1} \\ \cdots & \cdots & \cdots & \cdots \\ X_{r-1} \cdot X_1 & X_{r-1} \cdot X_2 & \cdots & X_{r-1} \cdot X_{r-1} \end{vmatrix}$$

To find the length of the projection of X_r onto the orthogonal complement of the space generated by $X_1, X_2, \cdots, X_{r-1}$, we write

(8) $\qquad X_r = Y + a_1 X_1 + a_2 X_2 + \cdots + a_{r-1} X_{r-1},$

where Y is the required projection, and therefore $Y \cdot X_i = 0$ for $i = 1, 2, \cdots, r-1$. It follows that

(9) $$Y \cdot Y = Y \cdot X_r,$$

and

(10) $$\begin{aligned} a_1 X_1 \cdot X_1 &+ a_2 X_1 \cdot X_2 + \cdots + a_{r-1} X_1 \cdot X_{r-1} = X_1 \cdot X_r, \\ a_1 X_2 \cdot X_1 &+ a_2 X_2 \cdot X_2 + \cdots + a_{r-1} X_2 \cdot X_{r-1} = X_2 \cdot X_r, \\ &\qquad \cdots \cdots \cdots \cdots \cdots \cdots \cdots \cdots \\ a_1 X_{r-1} \cdot X_1 &+ a_2 X_{r-1} \cdot X_2 + \cdots + a_{r-1} X_{r-1} \cdot X_{r-1} = X_{r-1} \cdot X_r. \end{aligned}$$

The determinant of the coefficients of equations 10 is b^2. Solving for $a_1, a_2, \cdots, a_{r-1}$, we get

$$b^2 a_1 = (-1)^{r-2} b_1,\ b^2 a_2 = (-1)^{r-3} b_2, \cdots, b^2 a_{r-1} = b_{r-1},$$

where $b_1, b_2, \cdots, b_{r-1}$ are the minors of the first $r-1$ elements in the last row of (7), and b^2 is the minor of the last element of this row. Substituting in (8), we get

$$\begin{aligned} b^2 Y &= b^2(-a_1 X_1 - a_2 X_2 - \cdots - a_{r-1} X_{r-1} + X_r) \\ &= (-1)^{r-1} b_1 X_1 + (-1)^{r-2} b_2 X_2 + \cdots + (-1) b_{r-1} X_{r-1} + b^2 X_r, \end{aligned}$$

whence from (9)

$$\begin{aligned} b^2 \|Y\|^2 &= b^2 Y \cdot X_r \\ &= (-1)^{r-1}[b_1 X_r \cdot X_1 - b_2 X_r \cdot X_2 + \cdots + (-1)^{r-1} b^2 X_r \cdot X_r]. \end{aligned}$$

The right-hand side of this equation is exactly the expansion of (7) by the elements of the last row, and the left-hand side is $b^2 \|Y\|^2$, the square of the r-volume of our r-dimensional parallelepiped. The theorem is therefore proved. We can also conclude that the determinant (7) is non-negative.

If a different set of $r-1$ of the vectors $X_1, X_2, \cdots, X_r$ had been chosen to define the base of the r-dimensional parallelepiped, the resulting expression for v^2 would have been the determinant (7) with a certain permutation applied to the subscripts on the X's. Owing to the fact that this determinant is symmetric about the main diagonal, the effect of such a permutation of subscripts is to permute the rows and columns of (7) in precisely the same manner. It follows that the determinant itself would be unchanged, and the volume is therefore independent of the base chosen.

COROLLARY. The n-volume of the n-dimensional parallelepiped in $\mathscr{V}_n(\mathscr{R})$ defined by the vectors

$$X_i = (x_{i1}, x_{i2}, \cdots, x_{in}) \qquad (i = 1, 2, \cdots, n)$$

is the absolute value of the determinant of the matrix

$$X = \begin{pmatrix} x_{11} & x_{12} & \cdots & x_{1n} \\ x_{21} & x_{22} & \cdots & x_{2n} \\ \cdot & \cdot & \cdot & \cdot \\ x_{n1} & x_{n2} & \cdots & x_{nn} \end{pmatrix}.$$

Proof. The determinant of XX^t is exactly the determinant (7) and is equal to the square of the determinant of X.

If the determinant X is positive, the vectors $X_1, X_2, \cdots, X_n$ are said to be *positively oriented*; if the determinant is negative, the vectors are said to be *negatively oriented*. Note that the orientation of a set of vectors depends on the order in which they are written.

Exercise 4.3

1. What is the number of vertices, edges, two-dimensional faces, and three-dimensional faces of a four-dimensional parallelepiped? Draw a plane projection of a four-dimensional cube.

2. How many r-dimensional faces has an n-dimensional parallelepiped?

3. Find the 3-volume of the three-dimensional parallelepipeds in $\mathscr{V}_4(\mathscr{R})$ defined by the vectors
    (a) $(2, 1, 0, -1), (3, -1, 5, 2), (0, 4, -1, 2)$.
    (b) $(1, 1, 0, 0), (0, 2, 2, 0), (0, 0, 3, 3)$.

4. Find the 2-volume of the parallelogram in $\mathscr{V}_4(\mathscr{R})$ two of whose edges are the vectors $(1, 3, -1, 6)$ and $(-1, 2, 4, 3)$.

5. Show that the parallelepiped in $\mathscr{V}_3(\mathscr{R})$ defined by the vectors $(2, 2, 1)$, $(1, -2, 2)$, and $(-2, 1, 2)$ is a cube. Find the volume of this cube.

6. Prove that, if the vectors $X_1, X_2, \cdots, X_r$ are mutually orthogonal, the r-volume of the parallelepiped defined by them is equal to the product of their lengths.

7. Prove that the vectors of any normal orthogonal basis of $\mathscr{V}_n(\mathscr{R})$ define a parallelepiped of unit volume.

8. Prove that r vectors $X_1, X_2, \cdots, X_r$ of $\mathscr{V}_n(\mathscr{R})$ are linearly dependent if and only if the determinant (7) is equal to zero.

9. Prove that two vectors X_1 and X_2 in the plane are positively oriented if and only if the sense of rotation from X_1 to X_2, through an angle less than $180°$, is counterclockwise.

CHAPTER FIVE

Transformations of Coordinates in a Vector Space

27. GENERAL COORDINATE SYSTEMS

We recall from Section 8, that, if V is any vector of a space $\mathscr{S}$ and if $E_1, E_2, \cdots, E_n$ is any basis of $\mathscr{S}$, we can write $V = x_1E_1 + x_2E_2 + \cdots + x_nE_n$. Once the basis is chosen, the vector V is uniquely determined by the ordered coefficients $(x_1, x_2, \cdots, x_n)$ of the basis elements, and these coefficients are called the coordinates of V relative to the E-basis. These coordinates are real numbers, if the field of scalars is $\mathscr{R}$, complex numbers if the field of scalars is $\mathscr{C}$. We shall call the vector $X = (x_1, x_2, \cdots, x_n)$ the *coordinate vector* of V relative to the E-basis. It is clear that, if kV is any scalar multiple of V, the coordinate vector of kV relative to the E-basis is kX, the same scalar multiple of X. Similarly, if V_1 and V_2 have coordinate vectors X_1 and X_2, then $V_1 + V_2$ has coordinate vector $X_1 + X_2$. Thus, as far as addition and scalar multiplication are concerned, the algebra of vectors V of $\mathscr{S}$ is faithfully reflected in the algebra of their coordinate vectors X. An important consequence of this is the following:

THEOREM 5.1. If $V_1, V_2, \cdots, V_r$ are any r vectors of $\mathscr{S}$ and $X_1, X_2, \cdots, X_r$ are their coordinate vectors relative to a basis $E_1, E_2, \cdots, E_n$, then $X_1, X_2, \cdots, X_r$ are linearly independent if and only if $V_1, V_2, \cdots, V_r$ are linearly independent.

Proof. If $X_i = (x_{i1}, x_{i2}, \cdots, x_{in})$, we have

$$V_1 = x_{11}E_1 + x_{12}E_2 + \cdots + x_{1n}E_n,$$
$$\cdots \cdots \cdots \cdots \cdots \cdots$$
$$V_r = x_{r1}E_1 + x_{r2}E_2 + \cdots + x_{rn}E_n,$$

and it is clear that any linear relation among $V_1, \cdots, V_r$ implies the same linear relation among the coefficients of $E_1, E_2, \cdots, E_n$ and therefore among the row vectors $X_1, X_2, \cdots, X_r$. Conversely a linear relation among $X_1, X_2, \cdots, X_r$ implies the same linear relation among $V_1, V_2, \cdots, V_r$.

In view of Theorem 5.1, in testing a set of vectors for linear dependence it is permissible to test instead their coordinate vectors relative to any basis.

28. TRANSFORMATION OF COORDINATES

We shall now investigate the relation between the coordinate vectors of a vector V relative to two different bases of $\mathscr{S}$. Let $E_1, \cdots, E_n$ and $F_1, \cdots, F_n$ be any such bases. Let the coordinate vectors of V be $X = (x_1, x_2, \cdots, x_n)$ and $Y = (y_1, y_2, \cdots, y_n)$ relative to the E-basis and F-basis, respectively. We know that the basis vectors are connected by equations of the form

$$E_1 = p_{11}F_1 + p_{12}F_2 + \cdots + p_{1n}F_n,$$
(1)
$$\cdots \cdots \cdots \cdots \cdots \cdots$$
$$E_n = p_{n1}F_1 + p_{n2}F_2 + \cdots + p_{nn}F_n,$$

where (by Problem 9, Exercise 2.2) the matrix $P = (p_{ij})$ is nonsingular. Now, since

$$V = x_1E_1 + x_2E_2 + \cdots + x_nE_n = y_1F_1 + y_2F_2 + \cdots + y_nF_n,$$

on substituting from (1) we get two expressions for V in terms of $F_1, \cdots, F_n$, and by Theorem 1.8 we may equate coefficients of $F_1, \cdots, F_n$ to get

$$y_1 = p_{11}x_1 + p_{21}x_2 + \cdots + p_{n1}x_n,$$
(2)
$$\cdots \cdots \cdots \cdots \cdots \cdots$$
$$y_n = p_{1n}x_1 + p_{2n}x_2 + \cdots + p_{nn}x_n.$$

Equations 2 give the coordinates of V relative to the F-basis in terms of the coordinates relative to the E-basis. These may be written in the compact matrix form $Y = XP$ which gives the relation between the two coordinate vectors.

The change from the E-basis to the F-basis described above is called a *transformation of coordinates* in $\mathscr{S}$. The matrix P of equations 1 is called the *matrix of this transformation of coordinates*. The matrix of equations 2 which gives the new coordinates of a vector in terms of the

Transformations of Coordinates 93

old is then P^t, and, if equations 2 are solved for the x's in terms of the y's, we see that the old coordinates are given in terms of the new by a set of linear equations with matrix $(P^t)^{-1}$. For purposes of reference we state these facts formally as

THEOREM 5.2. In a transformation of coordinates with matrix P the old basis vectors are given in terms of the new by linear equations with matrix P, the new coordinates of a vector are given in terms of the old by equations with matrix P^t, and the old coordinates are given in terms of the new by equations with matrix $(P^t)^{-1}$. Finally, in matrix notation, if X and Y are the coordinate vectors of V relative to the old and new bases, respectively, then $Y = XP$ and $Y^t = P^t X^t$.

We can also prove

THEOREM 5.3. Let $E_1, \cdots, E_n, F_1, \cdots, F_n, G_1, \cdots, G_n$ be three bases of $\mathscr{S}$. If the transformation from the E-basis to the F-basis has matrix P, and the transformation from the F-basis to the G-basis has matrix Q, then the transformation from the E-basis to the G-basis has matrix PQ.

Proof. If X, Y, and Z are the coordinate vectors of any vector V relative to the E-, F-, and G-bases, respectively, then by Theorem 5.2 $Y = XP$ and $Z = YQ = (XP)Q = X(PQ)$, and the result follows.

COROLLARY. If the transformation from the E-basis to the F-basis has matrix P, the transformation from the F-basis to the E-basis has matrix P^{-1}.

Proof. If in Theorem 5.3 we let $G_i = E_i$ ($i = 1, 2, \cdots, n$), we have $PQ = I$ and therefore $Q = P^{-1}$.

Exercise 5.1

1. Write the matrix of the following transformations of coordinates and the equations that give the old coordinates of a vector in terms of the new:

   (a) From the basis (1, 0), (0, 1) to the basis (2, 1), (3, 2) in $\mathscr{V}_2(\mathscr{R})$.

   (b) From the basis (− 1, 2), (2, 2) to the basis (− 1, 5), (− 1, − 2) in $\mathscr{V}_2(\mathscr{R})$.

   (c) From the basis (1, 0, 0), (0, 1, 0), (0, 0, 1) to the basis (2, − 1, 0), (1, 1, 1), (− 3, 0, 4) in $\mathscr{V}_3(\mathscr{R})$.

   (d) From the basis (1, 0, 1), (1, 1, 0), (0, 1, 1) to the basis (2, 1, 2), (3, − 1, 4), (1, 1, 2) in $\mathscr{V}_3(\mathscr{R})$.

2. Find the matrix of the transformation of coordinates from rectangular Cartesian coordinates (x, y) in the plane to oblique axes whose equations are $x - 2y = 0$ and $x + 2y = 0$ relative to the rectangular system. Assume that the unit of distance is the same in each system.

3. Find the coordinates of the vector (2, 1, 3, 4) of $\mathscr{V}_4(\mathscr{R})$ relative to the basis vectors $F_1 = (1, 1, 0, 0)$, $F_2 = (1, 0, 1, 0)$, $F_3 = (1, 0, 0, 1)$, $F_4 = (0, 0, 1, 1)$.

29. NORMAL ORTHOGONAL BASES. ORTHOGONAL MATRICES

In Section 10, it was shown that every real vector space has a normal orthogonal basis, that is to say, a basis consisting of mutually orthogonal unit vectors. In fact, by the Gram-Schmidt process a normal orthogonal basis can be constructed starting from an arbitrary unit vector as the first basis vector. In this section, we shall investigate the transformation of coordinates from one normal orthogonal basis to another and shall characterize the matrices of such transformation by simple algebraic properties.

Suppose first that $E_1, \cdots, E_n$ and $F_1, \cdots, F_n$ are two normal orthogonal bases of $\mathscr{V}_n(\mathscr{R})$ so that $E_i \cdot E_j = F_i \cdot F_j = \delta_{ij}$, where δ_{ij} is a Kronecker delta. Let P be the matrix of the transformation of coordinates from the E-basis to the F-basis so that the two bases are connected by equations 1. Forming the inner product $E_i \cdot E_j$, multiplying out the brackets, and using $F_i \cdot F_j = \delta_{ij}$, we find

$$E_i \cdot E_j = (p_{i1}F_1 + p_{i2}F_2 + \cdots + p_{in}F_n) \cdot (p_{j1}F_1 + p_{j2}F_2 + \cdots + p_{jn}F_n)$$
$$= p_{i1}p_{j1} + p_{i2}p_{j2} + \cdots + p_{in}p_{jn}$$
$$= \delta_{ij}.$$

Hence the matrix P of the transformation from one normal orthogonal basis to another has the property that its row vectors are mutually orthogonal unit vectors of $\mathscr{V}_n(\mathscr{R})$. Another way of stating this is by the matrix equation $PP^t = I$, since the element in the ith row and jth column of PP^t is the inner product of the ith and jth row vectors of P which is equal to δ_{ij}. However, $PP^t = I$ implies that $P^t = P^{-1}$ and $P^tP = I$. This tells us that the column vectors of P are also a set of n mutually orthogonal unit vectors.

DEFINITION. A real matrix P for which $PP^t = I$ is called an *orthogonal matrix*.

We have shown above that a necessary and sufficient condition that a matrix be orthogonal is that its row (and column) vectors form a mutually orthogonal set of unit vectors of $\mathscr{V}_n(\mathscr{R})$. It therefore follows that, if P is orthogonal, so also is P^t.

THEOREM 5.4. If $E_1, \cdots, E_n$ is a normal orthogonal basis of $\mathscr{V}_n(\mathscr{R})$ and if P is the matrix of the transformation from this basis to a basis

Transformations of Coordinates

$F_1, \cdots, F_n$, then a necessary and sufficient condition that the F-basis be normal orthogonal is that P be an orthogonal matrix.

Proof. We have already proved that, if $F_1, \cdots, F_n$ is a normal orthogonal basis, then P is orthogonal. Conversely, if P is orthogonal, then the matrix of the equations giving the F's in terms of the E's is P^{-1}, which is equal to P^t and hence is orthogonal. We therefore have

$$F_1 = p_{11}E_1 + p_{21}E_2 + \cdots + p_{n1}E_n,$$
$$\cdots \cdots \cdots \cdots \cdots$$
$$F_n = p_{1n}E_1 + p_{2n}E_2 + \cdots + p_{nn}E_n,$$

and, since $E_i \cdot E_j = \delta_{ij}$,

$$F_i \cdot F_j = p_{1i}p_{1j} + p_{2i}p_{2j} + \cdots + p_{ni}p_{nj} = \delta_{ij}.$$

Therefore $F_1, \cdots, F_n$ is a normal orthogonal basis.

Theorem 5.4 has an important consequence. If $X = (x_1, x_2, \cdots, x_n)$ and $Y = (y_1, y_2, \cdots, y_n)$ are any two vectors of $\mathscr{V}_n(\mathscr{R})$, it will be recalled that the inner product $X \cdot Y$ was defined by

$$X \cdot Y = x_1y_1 + x_2y_2 + \cdots + x_ny_n.$$

Thus the inner product of two vectors was defined in terms of their coordinates relative to a *particular* normal orthogonal basis, namely, $E_1 = (1, 0, \cdots, 0), E_2 = (0, 1, 0, \cdots, 0), \cdots, E_n = (0, 0, \cdots, 1)$. Now, if $F_1, \cdots, F_n$ is any other normal orthogonal basis and if P is the orthogonal matrix of the transformation of coordinates from the E- to the F-basis, we have by Theorem 5.2 that the coordinates X' and Y' of X and Y relative to the F-basis are given by $X' = XP$ and $Y' = YP$. We therefore have

$$X' \cdot Y' = x_1'y_1' + x_2'y_2' + \cdots + x_n'y_n' = X'Y'^t = XP(YP)^t = XPP^tY^t$$
$$= XY^t = X \cdot Y,$$

since $PP^t = I$. This shows that, in defining the inner product $X \cdot Y$ as above, the coordinates of X and Y relative to *any* normal orthogonal basis may be used. We state the result as follows:

THEOREM 5.5. *If V_1 and V_2 are any two vectors of $\mathscr{V}_n(\mathscr{R})$ and if $X = (x_1, x_2, \cdots, x_n)$, $Y = (y_1, y_2, \cdots, y_n)$ are their coordinate vectors relative to any normal orthogonal basis, then*

$$V_1 \cdot V_2 = XY^t = x_1y_1 + x_2y_2 + \cdots + x_ny_n = X \cdot Y,$$

this expression being independent of the particular normal orthogonal basis chosen.

COROLLARY. If two vectors of $\mathscr{V}_n(\mathscr{R})$ are orthogonal, their coordinate vectors relative to any normal orthogonal basis are also orthogonal.

The condition that a matrix P be orthogonal, namely, $PP^t = I$, may be written in a number of equivalent forms, for example, $P^t = P^{-1}$, $P^tP = I$, $(P^t)^{-1} = P$, $(P^{-1})^t = P$, etc. We now prove some other simple properties of orthogonal matrices.

THEOREM 5.6. *The inverse (or transpose) of an orthogonal matrix is orthogonal. The product of two or more orthogonal matrices is orthogonal. The determinant of an orthogonal matrix is equal to either 1 or* -1.

Proof. If P is orthogonal $P^t = P^{-1}$ and therefore $(P^t)^t = (P^{-1})^t = (P^t)^{-1}$ by Theorem 3.5. Hence P^t (and therefore P^{-1}) are orthogonal. If P and Q are orthogonal, then $(PQ)^t = Q^tP^t = Q^{-1}P^{-1} = (PQ)^{-1}$, and therefore PQ is orthogonal. Finally, if P is orthogonal, $PP^t = I$, and therefore $|PP^t| = |P|\,|P^t| = |P|^2 = 1$, and $|P| = \pm 1$.

30. ROTATION AND REFLECTION OF COORDINATE AXES

DEFINITION. An orthogonal matrix whose determinant is equal to 1 is called a *proper orthogonal matrix*, and one whose determinant is -1 is called an *improper* orthogonal matrix.

It is obvious that the inverse of an orthogonal matrix P is proper or improper according as P is proper or improper. Also the product of two proper or two improper orthogonal matrices is proper orthogonal, but the product of a proper and an improper orthogonal matrix is improper. A transformation of coordinates from one orthogonal basis to another is called proper orthogonal if its matrix is proper orthogonal.

THEOREM 5.7. *Two normal orthogonal bases of* $\mathscr{V}_n(\mathscr{R})$ *have the same orientation if and only if the transformation of coordinates from one basis to the other is proper orthogonal.*

Proof. Let

$$E_i = (e_{i1}, e_{i2}, \cdots, e_{in})$$
$$F_i = (f_{i1}, f_{i2}, \cdots, f_{in})$$
$(i = 1, 2, \cdots, n)$

be two normal orthogonal bases of $\mathscr{V}_n(\mathscr{R})$. The E-basis is positively or negatively oriented according as the determinant $|e_{ij}|$ is positive or negative. Let P be the matrix of the transformation from the E-basis to the F-basis so that

$$E_i = p_{i1}F_1 + p_{i2}F_2 + \cdots + p_{in}F_n \qquad (i = 1, 2, \cdots, n).$$

Transformations of Coordinates

Equating corresponding coordinates on each side of this equation, we get

$$e_{ij} = p_{i1}f_{1j} + p_{i2}f_{2j} + \cdots + p_{in}f_{nj} \qquad (i,j = 1, 2, \cdots, n),$$

and therefore $E = PF$, where $E = (e_{ij})$ and $F = (f_{ij})$. It follows that $|E| = |P|\,|F|$, and therefore the determinants $|E|$ and $|F|$ have like signs if $|P| = 1$ and opposite signs if $|P| = -1$. Hence a proper orthogonal transformation of coordinates preserves the orientation of the basis vectors, whereas an improper orthogonal transformation changes their orientation.

DEFINITION. A proper orthogonal transformation of coordinates from one normal orthogonal basis to another is called a *rotation* of the basis vectors or a *rotation of coordinate axes*.

In order to justify this definition of a rotation of axes in n-space, we must show that it corresponds in case $n = 3$ with our usual concept of a rotation of axes in space. A rotation of axes in three-dimensional space is usually defined as a continuous rigid motion of the coordinate axes which leaves the origin fixed. It is easy to see that such a transformation cannot change the orientation of the basis vectors. For let

$$E_i = (e_{i1}, e_{i2}, e_{i3}) \qquad (i = 1, 2, 3)$$

be a normal orthogonal basis for $\mathscr{V}_3(\mathscr{R})$ which can be transformed by a rotation into a new normal orthogonal basis $F_i = (f_{i1}, f_{i2}, f_{i3})$. Let $X_i = (x_{i1}, x_{i2}, x_{i3})$ be the coordinates of the basis vectors at any intermediate position so that each x_{ij} varies continuously from e_{ij} to f_{ij}. Since the determinant $|x_{ij}|$ is a continuous function of the coordinates x_{ij}, it varies continuously and therefore assumes every value between $|e_{ij}|$ and $|f_{ij}|$. Hence, if $|e_{ij}|$ and $|f_{ij}|$ differ in sign, $|x_{ij}|$ must assume the value zero at some intermediate position in the course of the rotation from the E-basis to the F-basis. This is impossible since in a rigid motion the basis vectors remain orthogonal and therefore linearly independent. Thus the orientation of the basis vectors cannot change in a continuous deformation of the basis vectors unless at some stage one of the basis vectors passes through the plane of the other two. A rotation of axes is therefore a transformation from one normal orthogonal basis to a similarly oriented normal orthogonal basis; in other words, it is a proper orthogonal transformation of coordinates.

Conversely by Theorems 5.4 and 5.7 every proper orthogonal transformation of coordinates in $\mathscr{V}_3(\mathscr{R})$ is a transformation from a normal orthogonal basis E_1, E_2, E_3 to a similarly oriented normal orthogonal basis F_1, F_2, F_3. We shall show that any such transformation is a rotation of axes. It is clear that E_1 and E_2 can be brought into coincidence with F_1 and F_2 by a continuous rigid motion of $E_1, E_2,$ and E_3, and that E_3 must then coincide either with F_3 or with $-F_3$. But, since F_1, F_2, F_3 have the same orientation as E_1, E_2, E_3, it follows that $F_1, F_2, -F_3$ have the opposite orientation. Hence, since a continuous rigid motion cannot change the orientation, the rigid motion that brings E_1 and E_2 into coincidence with F_1 and F_2 must also bring E_3 into coincidence with F_3. Hence every proper orthogonal transformation of coordinates is a rotation of axes. In $\mathscr{V}_3(\mathscr{R})$, therefore, the notions of proper orthogonal transformation of coordinates and rotation of coordinate axes are identical, and it is therefore natural to define a rotation of axes in $\mathscr{V}_n(\mathscr{R})$ to be a proper orthogonal transformation of coordinates.

Every improper orthogonal matrix P can clearly be converted into a proper orthogonal matrix R by changing the sign of all elements in the first column, which is equivalent to multiplying it on the right by the improper orthogonal matrix

$$Q = \begin{pmatrix} -1 & 0 & \cdots & 0 \\ 0 & 1 & \cdots & 0 \\ \cdot & \cdot & \cdot & \cdot \\ 0 & 0 & \cdots & 1 \end{pmatrix}.$$

This is the matrix of the transformation from the basis $F_1, F_2, \cdots, F_n$ to the basis $-F_1, F_2, \cdots, F_n$, and may be thought of as a reflection of the F_1 basis vector (or coordinate axis) in the hyperplane generated by $F_2, F_3, \cdots, F_n$ (for example, reflection of the x-axis in the yz-plane). We have then, $PQ = R$ or, since $Q^2 = 1$, $P = RQ$. Hence the transformation of coordinates with matrix P is equivalent to a rotation of axes with matrix R followed by a reflection of one coordinate axis in the hyperplane determined by the other $n-1$ coordinate axes.

Exercise 5.2

1. Find a normal orthogonal basis of $\mathscr{V}_3(\mathscr{R})$ of which $(\tfrac{1}{3}, \tfrac{2}{3}, \tfrac{2}{3})$ is the first basis vector, and write the matrix of the transformation of coordinates from the basis $(1, 0, 0), (0, 1, 0), (0, 0, 1)$ to this new basis.

2. Prove that every proper orthogonal second-order matrix has the form

$$\begin{pmatrix} \cos\theta & -\sin\theta \\ \sin\theta & \cos\theta \end{pmatrix}$$

Transformations of Coordinates

and every improper orthogonal second order matrix has the form

$$\begin{pmatrix} \cos\theta & \sin\theta \\ \sin\theta & -\cos\theta \end{pmatrix},$$

where θ is a suitably chosen angle.

3. Verify that the transformations of coordinates from the basis $E_1 = (1, 0, 0)$, $E_2 = (0, 1, 0)$, and $E_3 = (0, 0, 1)$ to each of the following bases is orthogonal, and write the matrix of the transformation in each case.

   (a) $F_1 = (\frac{2}{3}, \frac{2}{3}, \frac{1}{3})$, $F_2 = (-\frac{2}{3}, \frac{1}{3}, \frac{2}{3})$, $F_3 = (\frac{1}{3}, -\frac{2}{3}, \frac{2}{3})$.

   (b) $G_1 = (\frac{2}{7}, \frac{3}{7}, \frac{6}{7})$, $G_2 = (\frac{6}{7}, \frac{2}{7}, -\frac{3}{7})$, $G_3 = (\frac{3}{7}, -\frac{6}{7}, \frac{2}{7})$.

   (c) $H_1 = (\frac{3}{5}, 0, -\frac{4}{5})$, $H_2 = (\frac{4}{5}, 0, \frac{3}{5})$, $H_3 = (0, 1, 0)$.

4. Which of the three transformations in Problem 3 are rotations?

5. Referring to the vectors given in Problem 3, find the matrix of the transformation from the (a) G-basis to the E-basis, (b) E-basis to the H-basis, (c) G-basis to the H-basis, (d) F-basis to the G-basis.

6. Verify that all the transformations in Problem 5 are orthogonal. Which of them are rotations?

7. If $X = (x_1, x_2, x_3)$, $Y = (y_1, y_2, y_3)$ are any two vectors of $\mathscr{V}_3(\mathscr{R})$, their vector product is the vector $X \times Y = (x_2y_3 - x_3y_2, x_3y_1 - x_1y_3, x_1y_2 - x_2y_1)$. Prove that X, Y, $X \times Y$ are always positively oriented.

8. Generalize Theorem 5.7 as follows: Any two bases of $\mathscr{V}_n(\mathscr{R})$ are similarly oriented if and only if the matrix of transformation from one of these bases to the other has positive determinant. Deduce that it is impossible to continuously deform a positively oriented set of basis vectors into a negatively oriented set without passing one of these vectors through the space generated by the remaining ones. Interpret this result in two- and three-dimensional space.

CHAPTER SIX

Linear Transformations in a Vector Space

31. DEFINITION OF A LINEAR TRANSFORMATION AND ITS ASSOCIATED MATRIX RELATIVE TO A GIVEN BASIS

The student is probably familiar with the idea that a real function $f(x)$ defines a *mapping* of the real numbers into themselves in the sense that, given a real number x, a real number $f(x)$ is associated with it by means of the function f. The number $f(x)$ is uniquely determined by x, and may be thought of as the *image* of x under the mapping f. In this chapter we shall discuss similar mappings of a vector space $\mathscr{S}$ into itself. The type of mapping considered, however, will be much more restricted than the real functions that are usually studied in calculus. We shall consider only *single-valued linear* mappings of $\mathscr{S}$, usually called linear transformations. These are defined as follows:

DEFINITION. A *linear transformation* in a vector space $\mathscr{S}$ is a mapping of the space into itself which satisfies the following three conditions:

(a) Every vector V of $\mathscr{S}$ is mapped onto a uniquely determined vector V' of $\mathscr{S}$.

(b) If V is mapped onto V', then kV is mapped onto kV' for every scalar k.

(c) If V_1 is mapped onto V_1' and V_2 onto V_2', then $V_1 + V_2$ is mapped onto $V_1' + V_2'$.

An immediate consequence of (b) (on putting $k = -1$) is that, if V is mapped by a linear transformation onto V', then $-V$ is mapped onto $-V'$. It then follows from (c) that $V + (-V)$ is mapped onto $V' + (-V')$. Hence a linear transformation always maps the zero vector onto itself. Any vector V for which $V' = V$ is said to be *invariant*

Linear Transformations

under the linear transformation. The zero vector, therefore, is invariant under every linear transformation.

The vector V' onto which V is mapped by a linear transformation is called the *image* of V under the transformation, and V is called the *antecedent* of V'. Note that, although condition (a) of the definition ensures that the image V' is uniquely determined by V (i.e., that the mapping is single-valued), on the other hand, V is not necessarily uniquely determined by V'. A given vector may be the image of more than one antecedent or may not be an image at all. Conditions (b) and (c) of the definition are called the *linearity conditions*. They enable us to prove the fundamental properties of linear transformations contained in the next two theorems.

THEOREM 6.1. If a vector space $\mathscr{S}$ is subjected to a linear transformation, then

(a) the images of all vectors of any subspace $\mathscr{T}$ constitute a subspace $\mathscr{T}'$ of $\mathscr{S}$, called the *image space* of $\mathscr{T}$;

(b) the set of all antecedents of all vectors of a subspace $\mathscr{T}'$ of $\mathscr{S}$ constitutes a subspace $\mathscr{T}$ of $\mathscr{S}$.

Proof. (a) By conditions (b) and (c) of the definition of a linear transformation, if V_1' and V_2' belong to $\mathscr{T}'$ and are therefore images of vectors V_1, V_2 of $\mathscr{T}$, then kV_1' and $V_1' + V_2'$ are, respectively, the images of the vectors kV_1 and $V_1 + V_2$ of $\mathscr{T}$. Hence $\mathscr{T}'$ is closed under addition and scalar multiplication and is therefore a vector space.

(b) Let $\mathscr{T}'$ be a subspace of $\mathscr{S}$ and let $\mathscr{T}$ be the set of all antecedents of vectors of $\mathscr{T}'$. If $V \in \mathscr{T}$ then $V' \in \mathscr{T}'$, and therefore $kV' = (kV)' \in \mathscr{T}'$ and consequently $kV \in \mathscr{T}$. Similarly, if V_1 and V_2 belong to $\mathscr{T}$, V_1' and V_2' and therefore $V_1' + V_2'$ belong to $\mathscr{T}'$. But $V_1' + V_2' = (V_1 + V_2)'$, and therefore $V_1 + V_2 \in \mathscr{T}$. Hence $\mathscr{T}$ is a vector space.

Let $\mathscr{T}'$ be the image space of $\mathscr{T}$ under a linear transformation τ. If $\mathscr{T}' \subseteq \mathscr{U}$, we say that $\mathscr{T}$ is mapped *into* the space $\mathscr{U}$ by τ. If, however, $\mathscr{T}' = \mathscr{U}$, we say that $\mathscr{T}$ is mapped *onto* $\mathscr{U}$ by τ. To say that $\mathscr{T}$ is mapped into $\mathscr{U}$ does not exclude the possibility that the mapping is actually onto $\mathscr{U}$. It should also be noted that $\mathscr{T}$ need not be the space of *all* antecedents of vectors of $\mathscr{T}'$ although $\mathscr{T}$ must, of course, be contained in this space. If $\mathscr{T}' \subseteq \mathscr{T}$, then $\mathscr{T}$ is said to be *invariant* under the transformation τ. This does not necessarily imply that every vector of $\mathscr{T}$ is invariant but only that its image is in $\mathscr{T}$.

COROLLARY. The set of all vectors of $\mathscr{S}$ that are mapped by a linear transformation on the zero vector constitute a subspace of $\mathscr{S}$.

※ DEFINITION. The subspace of all vectors of $\mathscr{S}$ that are mapped by a linear transformation onto the zero vector is called the *kernel* of the transformation.

A second important consequence of conditions (b) and (c) in the definition of a linear transformation is that the image of every vector of the space $\mathscr{S}$ is completely determined when the images of any set of basis vectors are known. For suppose that $E_1, E_2, \cdots, E_n$ is a basis of $\mathscr{S}$ and that these vectors are mapped by a linear transformation onto E_1', $E_2', \cdots, E_n'$, respectively. It follows from (b) and (c) that the image of the vector $V = x_1 E_1 + x_2 E_2 + \cdots + x_n E_n$ is

(1) $$V' = x_1 E_1' + x_2 E_2' + \cdots + x_n E_n'.$$

Moreover, if any n vectors $E_1', \cdots, E_n'$ are chosen arbitrarily to be the images of the basis vectors, then the mapping $V \to V'$, where V' is defined by (1), is a linear transformation of $\mathscr{S}$ since the three conditions (a), (b), and (c) are clearly satisfied. We have therefore proved

THEOREM 6.2. *If $\mathscr{S}$ is a vector space of dimension n and if $E_1, E_2, \cdots, E_n$ are any linearly independent vectors of $\mathscr{S}$, there exists a uniquely determined linear transformation of $\mathscr{S}$ that maps E_i into E_i', where $E_1', E_2', \cdots, E_n'$ are any n arbitrarily chosen (not necessarily distinct) vectors of $\mathscr{S}$.*

Theorem 6.2 determines all linear transformations of $\mathscr{S}$. Briefly every linear transformation of S is uniquely determined when the images of a set of basis vectors are known, and conversely any choice of these images defines a linear transformation.

Now suppose that a space $\mathscr{S}$ with basis $E_1, E_2, \cdots, E_n$ is subjected to a linear transformation τ which maps E_i onto E_i' ($i = 1, 2, \cdots, n$). Since $E_1, \cdots, E_n$ is a basis, the image vectors E_i' can be expressed in terms of the E_i by equations of the form

$$E_1' = a_{11} E_1 + a_{12} E_2 + \cdots + a_{1n} E_n,$$
(2) $$\cdots \cdots \cdots \cdots \cdots$$
$$E_n' = a_{n1} E_1 + a_{n2} E_2 + \cdots + a_{nn} E_n.$$

It follows that the transformation τ is completely determined by equations 2, and therefore by the basis $E_1, \cdots, E_n$ and the matrix

$$A = \begin{pmatrix} a_{11} & a_{12} & \cdots & a_{1n} \\ \cdot & \cdot & \cdot & \cdot \\ a_{n1} & a_{n2} & \cdots & a_{nn} \end{pmatrix}.$$

Linear Transformations 103

The matrix A is called the *matrix of the transformation τ relative to the basis* $E_1, \cdots, E_n$. It is the matrix of the equations that give the images of the basis vectors in terms of the basis vectors themselves. If a different basis is used a different matrix will in general result. The relation between the matrices of a transformation relative to two different bases will be discussed in Section 35.

The above discussion associates a fixed matrix with each linear transformation, once a basis of the space has been chosen. Conversely, since by Theorem 6.2 an arbitrary choice of the image vectors $E_1', \cdots, E_n'$ defines a linear transformation, it follows that every matrix of order n defines a unique linear transformation by means of equations 2. Hence, once a basis of the space has been chosen, there is a one-to-one correspondence between the linear transformations of the space and the set of all matrices of order n whose elements lie in the field of scalars.

Now let $V = x_1 E_1 + x_2 E_2 + \cdots + x_n E_n$ be any vector of $\mathscr{S}$, and let $X = (x_1, x_2, \cdots, x_n)$ be its coordinate vector relative to the E-basis. The coordinates of the image of V under a linear transformation τ can easily be found when the matrix A of τ relative to the E-basis is known. For V is mapped by τ onto $V' = x_1 E_1' + x_2 E_2' + \cdots + x_n E_n'$, where $E_1', \cdots, E_n'$ are given in terms of $E_1, \cdots, E_n$ by (2). On substituting for $E_1', \cdots, E_n'$ from (2) and collecting coefficients of $E_1, \cdots, E_n$, we get

$$V' = x_1' E_1 + x_2' E_2 + \cdots + x_n' E_n,$$

where the coordinates $x_1' \ldots x_n'$, are given by the equations

$$x_1' = a_{11} x_1 + a_{21} x_2 + \cdots + a_{n1} x_n,$$

(3) $$\cdots \cdots \cdots \cdots \cdots$$

$$x_n' = a_{1n} x_1 + a_{2n} x_2 + \cdots + a_{nn} x_n,$$

whose matrix is A^t. Equations 3 can be written in matrix form: $X'^t = A^t X^t$ or $X' = XA$, where $X' = (x_1', x_2', \cdots, x_n')$ is the coordinate vector of V' relative to the E-basis. We state this result as

Theorem 6.3. Let $E_1, \cdots, E_n$ be any basis of a space $\mathscr{S}$. If a linear transformation of $\mathscr{S}$ maps a vector V on V' and if X and X' are the coordinate vectors of V and V' relative to the E-basis, then $X' = XA$, where A is the matrix of the transformation relative to the E-basis.

We note also that, since every matrix A of order n defines a linear transformation when a basis of $\mathscr{S}$ has been chosen, any set of equations of the form of equations 3 may be used to define a linear transformation of the space, where $(x_1, x_2, \cdots, x_n)$ and $(x_1', x_2', \cdots, x_n')$ are understood to be coordinates of a vector and its image relative to a fixed basis. If the space $\mathscr{S}$ is $\mathscr{V}_n(\mathscr{R})$ or $\mathscr{V}_n(\mathscr{C})$ and if no other basis is specified, it will be assumed that $(x_1, x_2, \cdots, x_n)$ and $(x'_1, x'_2, \cdots, x'_n)$ are the coordinates of V and V' relative to the basis $(1, 0, \cdots, 0), (0, 1, 0, \cdots, 0), \cdots, (0, 0, \cdots, 1)$, so that $V = X$ and $V' = X'$. Linear transformations in $\mathscr{V}_n(\mathscr{R})$ or $\mathscr{V}_n(\mathscr{C})$ will frequently be specified by equations of the form (3) with this basis tacitly assumed. When a transformation is defined in this way it should be noted that its matrix is the *transpose* of the matrix of the coefficients of equations 3.

32. SINGULAR AND NONSINGULAR LINEAR TRANSFORMATIONS

If τ is a linear transformation of $\mathscr{S}$ with matrix A relative to the basis $E_1, \cdots, E_n$, then the image of every vector of $\mathscr{S}$ has the form

$$x_1 E'_1 + \cdots + x_n E'_n,$$

where $E'_1, \cdots, E'_n$ are the images of $E_1, \cdots, E_n$. Hence $E'_1, \cdots, E'_n$ constitute a generating system for the image space of $\mathscr{S}$. The image space is therefore the whole space $\mathscr{S}$ if and only if $E'_1, \cdots, E'_n$ are linearly independent, and therefore a basis of $\mathscr{S}$. By equations 2 and Problem 9, Exercise 2.2, $E'_1, \cdots, E'_n$ are linearly independent if and only if the matrix A is nonsingular. This suggests the following

DEFINITION. A linear transformation τ of a vector space $\mathscr{S}$ is said to be *singular* if the image space under τ is a proper subspace of $\mathscr{S}$, and nonsingular if this image space is $\mathscr{S}$ itself.

In view of the above remark we see that a linear transformation is singular or nonsingular according as its matrix A, relative to any given basis, is singular or nonsingular.

The kernel of a linear transformation consists of all vectors of $\mathscr{S}$ that are mapped on the zero vector and therefore, by equations 3, all vectors of $\mathscr{S}$ whose coordinates relative to the E-basis satisfy linear equations

$$a_{11}x_1 + a_{21}x_2 + \cdots + a_{n1}x_n = 0,$$
$$\cdots \cdots \cdots \cdots \cdots$$
$$a_{1n}x_1 + a_{2n}x_2 + \cdots + a_{nn}x_n = 0.$$

It follows from Theorem 2.4 that the kernel contains nonzero vectors if and only if the matrix A, and therefore the linear transformation τ, is singular. This is a second distinction between singular and nonsingular transformations. A third may be derived from it. If two vectors V_1

and V_2 of $\mathscr{S}$ are mapped by τ on the same vector V', then $V_1 - V_2$ is mapped on $V' - V' = O$. Hence $V_1 - V_2$ belongs to the kernel of τ. If τ is nonsingular, it follows, since the kernel consists of O only, that $V_1 = V_2$ and therefore only *one* vector of $\mathscr{S}$ is mapped by τ onto any given vector. Since the image space is $\mathscr{S}$ itself, it follows that a nonsingular transformation is a *one-to-one* mapping of the space $\mathscr{S}$ onto itself. This result also follows from equations 3 since, if the matrix A is nonsingular, these equations (by Theorem 2.8) uniquely determine the coordinates X when the coordinates X' are given. On the other hand, if τ is singular, there is a nonzero vector W whose image is O, and consequently, if V is an arbitrary vector of $\mathscr{S}$, V and $V + W$ have the same image, $V' = V' + O$. A singular transformation is therefore always a many-to-one mapping of $\mathscr{S}$ onto a proper subspace of itself. This result also follows from equations 3 and Theorem 2.8.

The foregoing results may now be stated formally as

THEOREM 6.4. A nonsingular linear transformation of the space $\mathscr{S}$ is characterized by any one of the following equivalent properties:

(a) Its matrix relative to any basis is nonsingular.

(b) It is a one-to-one mapping of $\mathscr{S}$ onto itself.

(c) Its kernel is the zero vector.

If a linear transformation fails to satisfy any one of these three conditions, it fails to satisfy them all and it is then singular.

33. EXAMPLES OF LINEAR TRANSFORMATIONS

1. Consider the space $\mathscr{V}_2(\mathscr{R})$ and the linear transformation defined by the equations

$$x' = 3x,$$

$$y' = y.$$

The vector (x, y) is mapped by this transformation onto the vector $(3x, y)$. Geometrically this may be thought of as a stretching of the plane to the right and left of the y-axis in a direction parallel to the x-axis so that the x-coordinate of every point is multiplied by three while its y-coordinate remains unchanged. This is a simple example of a type of linear transformation which we shall call a magnification. In order to define a magnification in general suppose that n linearly independent vectors $E_1, E_2, \cdots, E_n$ of a real n-dimensional space $\mathscr{S}$ are mapped by a linear transformation τ_1, onto positive scalar multiples of themselves so that the image of E_i is $E_i' = a_i E_i$, where $a_i > 0$ $(i = 1, 2, \cdots, n)$. Since

$E_1, E_2, \cdots, E_n$ is a basis of $\mathscr{S}$, the transformation τ_1 is completely determined by its effect on $E_1, E_2, \cdots, E_n$, and it is clear that relative to this basis the matrix of τ_1 is

$$\begin{pmatrix} a_1 & 0 & \cdots & 0 \\ 0 & a_2 & \cdots & 0 \\ \cdot & \cdot & \cdot & \cdot \\ 0 & 0 & \cdots & a_n \end{pmatrix}.$$

Such a matrix, in which all elements except those in the main diagonal are zero, is called a *diagonal matrix*. In our case, all the diagonal elements are positive and the corresponding linear transformation τ_1 will be called a *magnification*. Every vector in $\mathscr{S}$ that is a scalar multiple of E_i is mapped by τ_1 into a_i times itself. Of course, if $0 < a_i < 1$, the corresponding vector E_i is "compressed" rather than "magnified," but the term magnification will be used to cover both these cases.

2. Consider the linear transformation τ_2 of $\mathscr{V}_3(\mathscr{R})$ defined by the equations $x' = x, y' = y, z' = -z$. Thus τ_2 maps every vector (x, y, z) of $\mathscr{V}_3(\mathscr{R})$ into its reflection $(x, y, -z)$ in the xy-plane. Such a linear transformation is an example of a reflection. The matrix of τ_2 is clearly

$$\begin{pmatrix} 1 & 0 & 0 \\ 0 & 1 & 0 \\ 0 & 0 & -1 \end{pmatrix}.$$

In general any linear transformation of $\mathscr{V}_n(\mathscr{R})$ that leaves $n-1$ vectors of a normal orthogonal basis invariant but maps the nth onto its negative will be called a *reflection*.

3. Consider the linear transformation τ_3 of $\mathscr{V}_2(\mathscr{R})$ defined by the equations

$$x' = x + y,$$
$$y' = y.$$

Here the vector (x, y) is mapped into $(x + y, y)$. Thus vectors that lie along the x-axis are unchanged by τ_3, but the terminal point (x, y) of any other vector is shifted to the right if it is in the upper half-plane, and to the left if it is in the lower half-plane, by an amount equal to the absolute value of its y-coordinate. A linear transformation of this type is called a *shear*.

Some idea of the type of distortion to which the plane is subjected by such a transformation may be obtained by investigating its effect on the lines $y = mx$ and the concentric circles $x^2 + y^2 = r^2$. Since we have

$$x = x' - y',$$
$$y = y',$$

the line $y = mx$ becomes $y' = m(x' - y')$ or $y' = mx'/(1 + m)$ and the circle $x^2 + y^2 = r^2$ is transformed into $(x' - y')^2 + y'^2 = r^2$ or

$$x'^2 - 2x'y' + 2y'^2 = r^2.$$

This is an ellipse with its axes of symmetry oblique to the coordinate axes.

4. Consider the space $\mathscr{V}_2(\mathscr{R})$ and the transformation τ_4 which rotates every vector (x, y) through a fixed angle θ in a counterclockwise direction. If r is the length of (x, y) and α its inclination to the x-axis, we have $x = r \cos \alpha$, $y = r \sin \alpha$ and

$$x' = r \cos (\alpha + \theta) = r \cos \alpha \cos \theta - r \sin \alpha \sin \theta,$$
$$y' = r \sin (\alpha + \theta) = r \sin \alpha \cos \theta + r \cos \alpha \sin \theta.$$

Hence a rotation of $\mathscr{V}_2(\mathscr{R})$ through an angle θ is a linear transformation defined by the equations

$$x' = x \cos \theta - y \sin \theta,$$
$$y' = x \sin \theta + y \cos \theta.$$

Since the matrix of a linear transformation is the transpose of the matrix of the equations giving the coordinates of the image in terms of those of the original vector, the matrices of the transformation τ_3 and τ_4 are, respectively,

$$\begin{pmatrix} 1 & 0 \\ 1 & 1 \end{pmatrix} \quad \text{and} \quad \begin{pmatrix} \cos \theta & \sin \theta \\ -\sin \theta & \cos \theta \end{pmatrix}.$$

Note that all the examples considered so far are nonsingular transformations.

5. Examples of singular linear transformations may also be given. Consider the space $\mathscr{V}_3(\mathscr{R})$, and map every vector $V = (x, y, z)$ onto its projection $V' = (x, y, 0)$ in the xy-plane. The equations of this transformation are $x' = x$, $y' = y$, and $z' = 0$, and its matrix is

$$\begin{pmatrix} 1 & 0 & 0 \\ 0 & 1 & 0 \\ 0 & 0 & 0 \end{pmatrix}.$$

This transformation is clearly singular since its matrix is, and since it maps the space $\mathscr{V}_3(\mathscr{R})$ onto the proper subspace of all vectors in the xy-plane. Its kernel is the subspace of all vectors which are mapped on $(0, 0, 0)$, namely those of the form $(0, 0, z)$, which lie along the z-axis.

6. Example 5 may be generalized as follows. Let $\mathscr{S}$ be any subspace of $\mathscr{V}_n(\mathscr{R})$ and $\mathscr{T}$ its orthogonal complement in $\mathscr{V}_n(\mathscr{R})$. By Theorem 4.2, every vector V or $\mathscr{V}_n(\mathscr{R})$ can be written uniquely in the form $V = S + T$, where $S \in \mathscr{S}$ and $T \in \mathscr{T}$. The mapping $V \to S$ of every vector V onto its projection on the space $\mathscr{S}$ is then a linear transformation of $\mathscr{V}_n(\mathscr{R})$, for, if $V_1 = S_1 + T_1$, then $V + V_1 = (S + S_1) + (T + T_1)$ and $kV = kS + kT$ are the corresponding decompositions of $V + V_1$ and kV. Hence kV is mapped on kS and $V + V_1$ on $S + S_1$. This transformation is singular if $\mathscr{S}$ is a proper subspace of $\mathscr{V}_n(\mathscr{R})$.

The foregoing examples will give the student some idea of the geometric nature of linear transformations, at least in spaces of two or three dimensions. Later the concept of a rotation will be extended to n-dimensional spaces, and we shall be able to show that every nonsingular linear transformation of $\mathscr{V}_n(\mathscr{R})$ is actually equivalent either to a reflection followed by a rotation and a magnification, or to a rotation followed by a magnification.

Exercise 6.1

1. Describe the geometric effect of the following linear transformations of the xy-plane, assuming that $a, b, c, k,$ are all constants different from zero. Find the equations of the images under each transformation of the lines $y = mx$ and the circles $x^2 + y^2 = r^2$.

(a) $x' = x,$
    $y' = ky.$

(b) $x' = ax,$
    $y' = by.$

(c) $x' = ax,$
    $y' = ay.$

(d) $x' = x,$
    $y' = cx + y.$

(e) $x' = \dfrac{x - y}{\sqrt{2}},$
    $y' = \dfrac{x + y}{\sqrt{2}}.$

(f) $x' = 3x + y,$
    $y' = -x + 3y.$

2. Describe the geometric effect of the following singular linear transformations, and in each case find the image space and the kernel of the transformation.

(a) $x' = 2x + y,$
    $y' = 6x + 3y.$

(b) $x' = x + y,$
    $y' = y + z,$
    $z' = x + 2y + z.$

3. Find a necessary and sufficient condition that the linear transformation

$$x' = ax + by,$$
$$y' = cx + dy$$

leave some nonzero vector (x, y) invariant.

4. Find a necessary and sufficient condition that the transformation of Problem 3 leave some one-dimensional subspace of $\mathscr{V}_2(\mathscr{R})$ invariant.

5. Find a linear transformation of the space $\mathscr{V}_2(\mathscr{R})$ that will map the circle $x^2 + y^2 = 1$ into the ellipse $x^2/4 + y^2/9 = 1$.

6. Find a linear transformation of $\mathscr{V}_2(\mathscr{R})$ that will map the circle $x^2 + y^2 = 1$ into the ellipse $13x^2 + 10xy + 13y^2 = 1$.

7. If X_1' and X_2' are the images of the vectors X_1 and X_2 of $\mathscr{V}_2(\mathscr{R})$ under the nonsingular linear transformation

$$X_1' = aX_1 + bX_2,$$
$$X_2' = cX_1 + dX_2,$$

prove that the area of the parallelogram defined by X_1' and X_2' is equal to that of the parallelogram defined by X_1 and X_2 multiplied by the absolute value of the determinant

$$\begin{vmatrix} a & b \\ c & d \end{vmatrix}$$

8. Prove that any bounded region R of the plane $\mathscr{V}_2(\mathscr{R})$, that is bounded by continuous curves, is mapped by the transformation of Problem 7 into a region R', whose area is equal to that of R multiplied by the absolute value of the determinant of the transformation. *Hint:* prove first for rectangular areas. Then express the area R as the limit of a sum of rectangular areas by means of an integral.

9. Deduce from Problem 8 that a linear transformation of $\mathscr{V}_2(\mathscr{R})$ always maps two regions of equal area into two regions of equal area.

10. What is a necessary and sufficient condition that a linear transformation of $\mathscr{V}_2(\mathscr{R})$ map every region R of the plane into a region having the same area as R?

11. Extend the results of Problems 7 through 10 to three-dimensional space by considering linear transformations of $\mathscr{V}_3(\mathscr{R})$.

12. Use the result of Problems 8 to find the area of the ellipse $x^2/a^2 + y^2/b^2 = 1$. *Hint:* apply a linear transformation to the xy plane which will map the circle $x^2 + y^2 = 1$ into the given ellipse.

13. Use the three-dimensional analogue of Problem 8 to find the volume of the ellipsoid $x^2/a^2 + y^2/b^2 + z^2/c^2 = 1$.

34. PROPERTIES OF NONSINGULAR LINEAR TRANSFORMATIONS

Although the matrix of a linear transformation is uniquely defined only if a fixed basis of the space is designated, we shall usually omit specific reference to the basis unless there is danger of ambiguity. Thus we shall speak of "the matrix of a linear transformation," assuming that a fixed basis is lurking in the background and that the matrices of different transformations are relative to the same basis unless otherwise stated.

Let σ and τ be any linear transformations of a space $\mathscr{S}$. If X^σ is the image of X under σ and $(X^\sigma)^\tau$ is the image of X^σ under τ, the reader may easily verify that the mapping ρ defined by $X \to (X^\sigma)^\tau$ is a linear transformation of $\mathscr{S}$. The transformation ρ defined in this way is called the *product* of σ and τ in the order named. We write $\rho = \sigma\tau$, or $X^{\sigma\tau} = (X^\sigma)^\tau$.

THEOREM 6.5. Let σ and τ be linear transformations of $\mathscr{S}$. If A_σ, A_τ, and $A_{\sigma\tau}$ are the matrices, relative to a basis (E_i), of σ, τ, and $\sigma\tau$, respectively, then $A_{\sigma\tau} = A_\sigma A_\tau$.

Proof. If a vector V has coordinate vector X relative to the basis (E_i), it follows from Theorem 6.3 that V^σ and $V^{\sigma\tau}$ have coordinate vectors XA_σ and $(XA_\sigma)A_\tau$, respectively. Since $(XA_\sigma)A_\tau = X(A_\sigma A_\tau)$, we have, again by Theorem 6.3, that $A_{\sigma\tau} = A_\sigma A_\tau$.

The student should be warned that some writers use the notation $\sigma\tau$ for the transformation defined by applying *first* τ and *then* σ. Since this converts Theorem 6.5 into the less desirable form $A_{\sigma\tau} = A_\tau A_\sigma$, the matrix of a transformation σ is then defined to be the transpose of the matrix that we have called A_σ. Since $A_{\sigma\tau}^t = (A_\tau A_\sigma)^t = A_\sigma^t A_\tau^t$, Theorem 6.5 is then restored to the form given above. In consulting reference books the student should first ascertain which notation the author has adopted. Since both are widely used it is important to be able to read books written in either notation.

The idea of the product of two transformations will be illustrated by an example. Let τ_1 be the linear transformation of $\mathscr{V}_2(\mathscr{R})$ defined by the equations $x' = 2x, y' = 5y$, and let τ_2 be that defined by

$$x' = (x - y)/\sqrt{2},$$
$$y' = (x + y)/\sqrt{2}.$$

The matrices of τ_1 and τ_2 are, respectively,

$$A_1 = \begin{pmatrix} 2 & 0 \\ 0 & 5 \end{pmatrix} \quad \text{and} \quad A_2 = \begin{pmatrix} \dfrac{1}{\sqrt{2}} & \dfrac{1}{\sqrt{2}} \\ -\dfrac{1}{\sqrt{2}} & \dfrac{1}{\sqrt{2}} \end{pmatrix}$$

If we apply first τ_1, the vector (x, y) is mapped onto $(x', y') = (2x, 5y)$. Now τ_2 maps (x', y') onto

$$(x'', y'') = \left(\frac{x' - y'}{\sqrt{2}}, \frac{x' + y'}{\sqrt{2}} \right) = \left(\frac{2x - 5y}{\sqrt{2}}, \frac{2x + 5y}{\sqrt{2}} \right)$$

Linear Transformations

The equations of the product transformation $\tau_1\tau_2$ are therefore

$$x'' = \frac{2x - 5y}{\sqrt{2}}$$

$$y'' = \frac{2x + 5y}{\sqrt{2}}$$

and its matrix is

$$\begin{pmatrix} \dfrac{2}{\sqrt{2}} & \dfrac{2}{\sqrt{2}} \\ -\dfrac{5}{\sqrt{2}} & \dfrac{5}{\sqrt{2}} \end{pmatrix} = A_1 A_2.$$

Note that the product $\tau_2\tau_1$ is an entirely different transformation, namely,

$$x'' = \frac{2(x-y)}{\sqrt{2}}$$

$$y'' = \frac{5(x+y)}{\sqrt{2}}$$

with matrix

$$\begin{pmatrix} \sqrt{2} & \dfrac{5}{\sqrt{2}} \\ -\sqrt{2} & \dfrac{5}{\sqrt{2}} \end{pmatrix} = A_2 A_1.$$

Geometrically the effect of τ_1 is a magnification by a factor 2 in the direction of the x-axis and a magnification by a factor 5 in the direction of the y-axis. The effect of τ_2 is a rotation through 45°. The student should satisfy himself that the combined effect on a vector of these two operations depends on the order in which they are performed.

If τ_1, τ_2, and τ_3 are any three linear transformations of $\mathscr{S}$ it follows from the definition of the product transformation that $\tau_1(\tau_2\tau_3)$ has the same effect on $\mathscr{S}$ as $(\tau_1\tau_2)\tau_3$. Hence $\tau_1(\tau_2\tau_3) = (\tau_1\tau_2)\tau_3$; that is, *multiplication of linear transformations is associative*. This also follows from the fact that multiplication of the corresponding matrices is associative.

The linear transformation whose matrix is the unit matrix evidently maps each basis vector on itself and therefore leaves every vector of the space invariant. It is called the *identity transformation* and is usually represented by the same symbol I as the unit matrix. It will be clear from the context whether I stands for the identity transformation or for the unit matrix. Note that the identity transformation is represented by the unit matrix relative to *every* basis of the space.

Suppose that τ is any nonsingular linear transformation of a space and that its matrix relative to the basis $E_1, \cdots, E_n$ is A. We know from Theorem 6.4 that A is nonsingular and that τ is a one-to-one mapping $V \to V'$ of the space onto itself. Since A is nonsingular, it has a unique inverse A^{-1} and $AA^{-1} = I$. It follows that there is a uniquely determined linear transformation τ^{-1} with matrix A^{-1} such that $\tau\tau^{-1} = I$, the identity transformation. Also, since $A^{-1}A = I$, we have also $\tau^{-1}\tau = I$. The transformation τ^{-1} is called the *inverse transformation* of τ. If V' is the image of any vector V under τ, then V is the image of V' under τ^{-1}. An inverse transformation can only be defined if τ is nonsingular since otherwise not every vector of $\mathscr{S}$ is an image under τ and such a vector would have no image under τ^{-1}. It is clear that τ^{-1} is nonsingular whenever it is defined since A^{-1} is nonsingular when A is. Moreover the product of nonsingular transformations is nonsingular since the product of nonsingular matrices is.

DEFINITION. A set $\mathscr{G}$ of nonsingular linear transformations of a space is called a *group* of linear transformations if (a) the product of any two transformations that belong to $\mathscr{G}$ also belongs to $\mathscr{G}$ and (b) the inverse of any transformation in $\mathscr{G}$ belongs to $\mathscr{G}$.

We see therefore that the set of all nonsingular transformations of a space $\mathscr{S}$ constitute a group in the above sense. This is called the *full linear group* on $\mathscr{S}$. The set of all linear transformations of $\mathscr{S}$ does not form a group since the singular transformations do not have inverses. There is a very extensive theory of groups of linear transformations which will not be considered in this book. Some of the more important of these groups, however, will be mentioned as we come to them.

THEOREM 6.6. *If vectors $V_1, V_2, \cdots, V_r$ of a space $\mathscr{S}$ are mapped by a nonsingular linear transformation τ onto the vectors $V_1', V_2', \cdots, V_r'$, then $V_1', V_2', \cdots, V_r'$ are linearly independent if and only if $V_1, V_2, \cdots, V_r$ are linearly independent.*

Proof. Relative to a fixed basis of $\mathscr{S}$, let the matrix of τ be A and the coordinate vectors of $V_1, \cdots, V_r$ and $V_1', \cdots, V_r'$ be $X_1, \cdots, X_r$ and $X_1', \cdots, X_r'$, respectively. Hence, by Theorem 6.3, $X_i' = X_iA$ ($i = 1, 2, \cdots, r$). Also by Theorem 5.1 it is sufficient to prove the required result for the coordinate vectors $X_1, \cdots, X_r$ and $X_1', \cdots, X_r'$. Now suppose that the vectors X_i' satisfy the linear relation

$$c_1X_1' + c_2X_2' + \cdots + c_rX_r' = O$$

so that

$$c_1(X_1A) + c_2(X_2A) + \cdots + c_r(X_rA) = O,$$

whence
$$(c_1X_1 + c_2X_2 + \cdots + c_rX_r)A = 0.$$

The last equation states that the vector $Y = c_1X_1 + c_2X_2 + \cdots + c_rX_r$ is a solution vector of the system of linear homogeneous equations $YA = O$ (or $A^tY^t = O$) whose matrix is A^t. Since A, and therefore A^t, is nonsingular, Theorem 2.4 tells us that $Y = O$. Hence every linear relation that holds among $X'_1, X'_2, \cdots, X'_r$ also holds among $X_1, X_2, \cdots, X_r$.

Conversely, if
$$c_1X_1 + c_2X_2 + \cdots + c_rX_r = O,$$
then
$$(c_1X_1 + c_2X_2 + \cdots + c_rX_r)A = O$$
and
$$c_1X_1A + c_2X_2A + \cdots + c_rX_rA = c_1X'_1 + c_2X'_2 + \cdots + c_rX'_r = O.$$

Consequently, the X_i's and the X'_i's satisfy precisely the same linear relations, and hence the X'_i's are linearly independent if and only if the X_i's are. This completes the proof.

COROLLARY. *Under a nonsingular linear transformation of a space $\mathscr{S}$ every subspace $\mathscr{T}$ of $\mathscr{S}$ is mapped onto an image space $\mathscr{T}'$ whose dimension is equal to that of $\mathscr{T}$.*

This follows since $\mathscr{T}'$ is generated by the images of the basis vectors of $\mathscr{T}$ and these images are linearly independent as a result of the theorem.

The following theorem generalizes the result stated in Problem 7, Exercise 6.1.

THEOREM 6.7. *If $X_1, X_2, \cdots, X_n$ are n linearly independent vectors of $\mathscr{V}_n(\mathscr{R})$ which are mapped by a nonsingular linear transformation τ, with matrix A, onto $X'_1, X'_2, \cdots, X'_n$, then the n-volume of the parallelepiped defined by $X'_1, X'_2, \cdots, X'_n$ is equal to that of the parallelepiped defined by $X_1, X_2, \cdots, X_n$ multiplied by the absolute value of the determinant of A.*

Proof. Let X be the matrix of which $X_1, X_2, \cdots, X_n$ are the row vectors and X' the matrix of which $X'_1, X'_2, \cdots, X'_n$ are the row vectors. Since $X_iA = X'_i$, we have $XA = X'$, and therefore $|X'| = |X| \, |A|$. But the volume of the parallelepiped defined by the X_i is the absolute value of the determinant $|X|$, whereas that of the parallelepiped defined by the X'_i is the absolute value of $|X'|$. The result therefore follows.

Exercise 6.2

1. Let τ_1 be the linear transformation of $\mathscr{V}_2(\mathscr{R})$ defined by the equations $x' = -y, y' = 2x$ and let τ_2 be that defined by

$$x' = (x - \sqrt{3}y)/2,$$
$$y' = (\sqrt{3}x + y)/2.$$

Find the transformations $\tau_1\tau_2, \tau_2\tau_1, \tau_1^2, \tau_2^2, \tau_2^3, \tau_2^{12}, \tau_1^{-1}$, and τ_2^{-1}. (Note that, if τ is a linear transformation, τ^r means the product $\tau\tau\tau \cdots$ to r factors, τ^{-r} means $(\tau^{-1})^r$, and τ^0 is defined to be I, the identity transformation.)

2. Prove that the following sets of linear transformations of $\mathscr{V}_n(\mathscr{R})$ form groups:

   (a) The set of all linear transformations whose matrices relative to a fixed basis have determinant ± 1.

   (b) The set of all linear transformations whose matrices have determinant equal to 1.

   (c) The set of all linear transformations whose matrices are orthogonal.

   (d) The set of all (positive, zero, and negative) powers of a fixed nonsingular linear transformation τ.

3. Prove that, if τ is a linear transformation of $\mathscr{V}_n(\mathscr{R})$ whose matrix has rank r, then τ maps $\mathscr{V}_n(\mathscr{R})$ onto a subspace of dimension r and that the kernel of τ has dimension $n - r$.

4. Let τ be the linear transformation that rotates every vector of $\mathscr{V}_2(\mathscr{R})$ counterclockwise through an acute angle whose tangent is $\frac{4}{3}$. Let σ be the linear transformation that multiplies the first coordinate of every vector of $\mathscr{V}_2(\mathscr{R})$ by 3. Find the matrices of $\sigma\tau$ and $\tau\sigma$ relative to the basis $E_1 = (1, 0)$, $E_2 = (0, 1)$.

5. If τ maps $E_1 = (1, 0)$ and $E_2 = (0, 1)$ onto the vectors $F_1 = (2, 5)$ and $F_2 = (-1, 6)$, respectively, find the matrices of τ and τ^{-1} relative to the basis E_1, E_2 and also relative to the basis F_1, F_2.

6. Suppose τ maps $F_1 = (2, -1)$ onto $G_1 = (4, 3)$ and $F_2 = (1, 3)$ onto $G_2 = (-1, -2)$. Find the matrix of τ relative to (a) the basis F_1, F_2; (b) the basis G_1, G_2; (c) the basis $E_1 = (1, 0)$, $E_2 = (0, 1)$.

35. THE MATRICES OF A LINEAR TRANSFORMATION RELATIVE TO DIFFERENT BASES

We have seen that a linear transformation of $\mathscr{S}$ is associated with a fixed matrix relative to any given basis of the space. We shall now investigate the relationship between the matrices of a given linear transformation relative to two different bases. Let $E_1, E_2, \cdots, E_n$ and

$F_1, F_2, \cdots, F_n$ be any two bases of the space $\mathscr{S}$, and let P be the matrix of the transformation of coordinates from the E-basis to the F-basis. Let V be any vector of the space, and let X and Y be its coordinate vectors relative to the E-basis and the F-basis, respectively. Now let τ be the linear transformation of $\mathscr{S}$ whose matrix relative to the E-basis is A, so that by Theorem 6.3 the image V' of V has coordinate vector $X' = XA$ relative to the E-basis. By Theorem 5.2 the coordinate vector of V' relative to the F-basis is $Y' = X'P$. Since, also by Theorem 5.2, $Y = XP$, and therefore $X = YP^{-1}$, we have

$$Y' = X'P = (XA)P = [(YP^{-1})A]P = Y(P^{-1}AP).$$

Hence the matrix that connects the F-coordinates of V and V' is $P^{-1}AP$, and this is the matrix of τ relative to the F-basis. We have therefore proved the following important result:

THEOREM 6.8. *If A is the matrix of a linear transformation τ of $\mathscr{S}$ relative to the basis $E_1, E_2, \cdots, E_n$, then the matrix of τ relative to the basis $F_1, F_2, \cdots, F_n$ is $P^{-1}AP$, where P is the matrix of transformation of coordinates from the E-basis to the F-basis.*

Since it is very important for what follows that the student understand the content of Theorem 6.8, we shall illustrate it with a numerical example. Consider the transformation τ of $\mathscr{V}_2(\mathscr{R})$ (i.e., of the xy-plane) defined by the equations

$$x' = 2x,$$
$$y' = 3y.$$

The matrix of τ relative to the basis $E_1 = (1, 0)$, $E_2 = (0, 1)$ is clearly

$$A = \begin{pmatrix} 2 & 0 \\ 0 & 3 \end{pmatrix},$$

since these basis vectors are mapped by τ onto $E_1' = (2, 0) = 2E_1$ and $E_2' = (0, 3) = 3E_2$. Now consider two new basis vectors $F_1 = (1, 2)$ and $F_2 = (3, -1)$ which are mapped by τ onto $F_1' = (2, 6)$ and $F_2' = (6, -3)$. By solving the necessary linear equations, we can express F_1', F_2' in terms of F_1, F_2 by the equations

(4)
$$F_1' = \tfrac{10}{7}F_1 - \tfrac{6}{7}F_2,$$
$$F_2' = -\tfrac{3}{7}F_1 + \tfrac{11}{7}F_2.$$

The linear transformation τ is equally well defined by equations 4 which give the images of the basis vectors F_1 and F_2. The matrix of τ relative to the F-basis is by definition

$$B = \begin{pmatrix} \tfrac{20}{7} & -\tfrac{3}{7} \\ -\tfrac{2}{7} & \tfrac{15}{7} \end{pmatrix}$$

Since

$$F_1 = E_1 + 2E_2,$$
$$F_2 = 3E_1 - E_2,$$

we get on solving for E_1, E_2

$$E_1 = \tfrac{1}{7}F_1 + \tfrac{3}{7}F_2,$$
$$E_2 = \tfrac{3}{7}F_1 - \tfrac{1}{7}F_2.$$

The matrix of the transformation of coordinates from the E-basis to the F-basis is therefore

$$P = \begin{pmatrix} \tfrac{1}{7} & \tfrac{3}{7} \\ \tfrac{3}{7} & -\tfrac{1}{7} \end{pmatrix}$$

and its inverse is

$$P^{-1} = \begin{pmatrix} 1 & 2 \\ 3 & -1 \end{pmatrix}$$

Hence

$$P^{-1}AP = \begin{pmatrix} 1 & 2 \\ 3 & -1 \end{pmatrix} \begin{pmatrix} 2 & 0 \\ 0 & 3 \end{pmatrix} \begin{pmatrix} \tfrac{1}{7} & \tfrac{3}{7} \\ \tfrac{3}{7} & -\tfrac{1}{7} \end{pmatrix} = \begin{pmatrix} 2 & 6 \\ 6 & -3 \end{pmatrix} \begin{pmatrix} \tfrac{1}{7} & \tfrac{3}{7} \\ \tfrac{3}{7} & -\tfrac{1}{7} \end{pmatrix}$$

$$= \begin{pmatrix} \tfrac{20}{7} & -\tfrac{3}{7} \\ -\tfrac{2}{7} & \tfrac{15}{7} \end{pmatrix} = B.$$

The result of Theorem 6.8 is therefore verified in this case.

We make a further remark concerning the above example. To say that the linear transformation τ has matrix

$$\begin{pmatrix} 2 & 0 \\ 0 & 3 \end{pmatrix}$$

relative to the basis $E_1 = (1, 0)$, $E_2 = (0, 1)$ at once reveals the geometric nature of τ, namely, that it is a magnification by a factor 2 in the direction of E_1 and by a factor 3 in the direction of E_2. But to say that the matrix of τ relative to the basis $F_1 = (1, 2)$, $F_2 = (3, -1)$ is

$$\begin{pmatrix} \tfrac{20}{7} & -\tfrac{3}{7} \\ -\tfrac{2}{7} & \tfrac{15}{7} \end{pmatrix}$$

Linear Transformations

reveals no information about the geometric nature of τ without further investigation. This illustrates the fact that, although, algebraically speaking, a linear transformation is completely defined when its matrix relative to any given basis is known, the geometric nature of the transformation is not usually revealed automatically by this matrix. The problem of determining the geometric effect on the space of a given linear transformation τ usually involves choosing a basis of the space in such a way that the matrix of τ relative to the chosen basis takes some particularly simple form which will reveal geometric properties of τ to us. Usually what is desired is a matrix in diagonal form, although this is not always possible.

The importance of Theorem 6.8 lies in the fact that it gives us an algebraic method of investigating the geometric effect of a given linear transformation. The given transformation τ is defined by its matrix A relative to a basis $E_1, E_2, \cdots, E_n$, but it is equally well defined by its matrix $P^{-1}AP$ relative to the basis $F_1, F_2, \cdots, F_n$, where P is the matrix of the transformation of coordinates from the E-basis to the F-basis. Our problem is therefore reduced to finding a nonsingular matrix P such that the matrix $P^{-1}AP$ gives us as much information as possible about the transformation τ. Much of the rest of this book is devoted to the problem of investigating various special types of linear transformations in this way. We shall be particularly interested in determining for exactly what matrices A a nonsingular matrix P can be found such that $P^{-1}AP$ is a diagonal matrix. The problem will not be completely solved in this book but only for a number of important special cases.

Exercise 6.3

1. If a linear transformation of $\mathscr{V}_2(\mathscr{R})$ maps $(1, 1)$ into $(2, -1)$ and $(2, 3)$ into $(-1, 3)$, find its matrix relative to the following bases:

   (a) $F_1 = (1, 0),\qquad F_2 = (0, 1)$.
   (b) $F_1 = (1, 1),\qquad F_2 = \overline{(2, -1)}$.
   (c) $F_1 = (-1, -1),\quad F_2 = (-2, 4)$.

2. Let τ be the linear transformation of $\mathscr{V}_2(\mathscr{R})$ whose matrix relative to the basis $(1, 0), (0, 1)$ is

$$\begin{pmatrix} -\frac{1}{2} & -\frac{5}{2} \\ -\frac{5}{2} & -\frac{1}{2} \end{pmatrix}.$$

Find the matrix of τ relative to a new basis obtained by rotating the original basis vectors through $45°$, and deduce from this new matrix the geometric effect of τ on the xy-plane.

3. A matrix A is said to be symmetric if $A^t = A$. Prove that, if τ is a linear transformation of $\mathscr{V}_n(\mathscr{R})$ whose matrix relative to a given normal orthogonal

basis is diagonal, then its matrix relative to any normal orthogonal basis is symmetric.

4. Prove that all the matrices of a linear transformation of $\mathscr{S}$ relative to different bases have the same determinant.

5. Find the matrix relative to the basis $E_1 = (1, 0)$, $E_2 = (0, 1)$ of the transformation τ of $\mathscr{V}_2(\mathscr{R})$ that magnifies every vector in the direction of $(1, 1)$ by a factor 6 and every vector in the direction of $(1, 3)$ by a factor 2. *Hint:* first write the matrix of τ relative to the basis $F_1 = (1, 1)$, $F_2 = (1, 3)$, and then transform to the E-basis.

6. Find the matrix, relative to the normal orthogonal basis

$$E_1 = (\tfrac{2}{3}, \tfrac{2}{3}, -\tfrac{1}{3}), \quad E_2 = (\tfrac{1}{3}, -\tfrac{2}{3}, -\tfrac{2}{3}), \quad E_3 = (\tfrac{2}{3}, -\tfrac{1}{3}, \tfrac{2}{3}),$$

of $\mathscr{V}_3(\mathscr{R})$, of the linear transformation τ defined by the equations

$$x' = 2x,$$
$$y' = 4y,$$
$$z' = 3z.$$

In these equations (x, y, z), (x', y', z') are understood to be coordinates relative to the basis $(1, 0, 0)$, $(0, 1, 0)$, and $(0, 0, 1)$.

7. How could you determine whether a given linear transformation

$$x' = ax + by,$$
$$y' = cx + dy$$

of $\mathscr{V}_2(\mathscr{R})$ is a magnification?

8. Which of the following linear transformations of $\mathscr{V}_2(\mathscr{R})$ are magnifications? For those that are, find the directions in which the magnification takes place, and the constant factor by which vectors in these directions are multiplied.

(a) $x' = 4x + y,$
$y' = 2x + 3y.$

(b) $x' = x + 3y,$
$y' = 3x - y.$

(c) $x' = 2x - y,$
$y' = 3x + 2y.$

36. ORTHOGONAL TRANSFORMATIONS

Let τ be a linear transformation in $\mathscr{V}_n(\mathscr{R})$, and denote by X' the image of the vector X under τ. The transformation τ is said to *preserve lengths* if, for every vector X in $\mathscr{V}_n(\mathscr{R})$, $\|X\| = \|X'\|$.

THEOREM 6.9. A necessary and sufficient condition that a linear transformation τ in $\mathscr{V}_n(\mathscr{R})$ preserve lengths is that τ preserve inner products; that is, for any two vectors X and Y in $\mathscr{V}_n(\mathscr{R})$, $X \cdot Y = X' \cdot Y'$ where X' and Y' are the images of X and Y under τ.

Proof. The condition is obviously sufficient, since, if inner products are preserved, $\|X\| = (X \cdot X)^{1/2} = (X' \cdot X')^{1/2} = \|X'\|$. The necessity follows from the identity

$$X \cdot Y = \tfrac{1}{2}\{\|X + Y\|^2 - \|X\|^2 - \|Y\|^2\},$$

Linear Transformations

which, together with the fact that $(X + Y)' = X' + Y'$, shows that, if all lengths are preserved, all inner products must also be preserved.

COROLLARY. *If the linear transformation τ in $\mathscr{V}_n(\mathscr{R})$ preserves lengths, it also preserves the angles between vectors; i.e. if X', Y' are the images under τ of X, Y, then the angle θ between X and Y is equal to the angle θ' between X' and Y'. In particular, if X and Y are orthogonal, so also are X' and Y'.*

Proof. Since $\cos \theta = \dfrac{X \cdot Y}{\|X\|\ \|Y\|} = \dfrac{X' \cdot Y'}{\|X'\|\ \|Y'\|} = \cos \theta'$, and since both θ and θ' are not greater than $180°$, we have $\theta = \theta'$. If $X \cdot Y = 0$, then $X' \cdot Y' = 0$, and hence orthogonality of X and Y implies that of X' and Y'.

THEOREM 6.10. *A necessary and sufficient condition that a linear transformation τ or $\mathscr{V}_n(\mathscr{R})$ preserve lengths is that its matrix A relative to the basis $E_1 = (1, 0, \cdots, 0), \cdots, E_n = (0, 0, \cdots, 1)$ be orthogonal.*

Proof. If A is orthogonal, $AA^t = I$, and we have

$$X' \cdot Y' = X'Y'^t = (XA)(YA)^t = XAA^tY^t = XY^t = X \cdot Y.$$

Hence τ preserves inner products, and therefore lengths.

Conversely, if $X' \cdot Y' = X \cdot Y$ for all vectors X and Y, we have

$$X' \cdot Y' = X'Y'^t = XAA^tY^t = XY^t$$

Now, if $AA^t = (c_{ij})$, the last equation implies that

$$\sum_{i=1}^n \sum_{j=1}^n c_{ij}x_iy_j = x_1y_1 + x_2y_2 + \cdots + x_ny_n$$

for *all* values of $x_1, \cdots, x_n, y_1, \cdots, y_n$. Putting $x_r = y_s = 1$ and $x_i = y_j = 0$ for $i \neq r$ and $j \neq s$, we see that $c_{ij} = \delta_{ij}$ and therefore $AA^t = (\delta_{ij}) = I$. Hence A is orthogonal as required.

COROLLARY. *The linear transformation τ of $\mathscr{V}_n(\mathscr{R})$ preserves lengths if and only if its matrix relative to any normal orthogonal basis is orthogonal.*

Proof. If $F_1, F_2, \cdots, F_n$ is any normal orthogonal basis, then, by Theorem 5.4, the matrix P of the transformation of coordinates from the E-basis to the F-basis is orthogonal. If A is the matrix of τ relative to the E-basis, then, by Theorem 6.8, $B = P^{-1}AP$ is the matrix of τ relative to the F-basis. By Theorem 5.6, B is orthogonal if A is, and also, since $A = PBP^{-1}$, A is orthogonal if B is. It follows that τ preserves lengths if and only if its matrix relative to the F-basis is orthogonal.

DEFINITION. A linear transformation of $\mathscr{V}_n(\mathscr{R})$ which preserves lengths is called an *orthogonal transformation*.

In view of Theorem 6.10 and its corollary an orthogonal transformation could also be defined as a linear transformation of $\mathscr{V}_n(\mathscr{R})$ whose matrix relative to any (and therefore every!) normal orthogonal basis is orthogonal.

An orthogonal transformation of $\mathscr{V}_n(\mathscr{R})$ with matrix A relative to any given normal orthogonal basis is said to be *proper* or *improper* according as $|A| = 1$ or $|A| = -1$. These properties also are invariant under transformation of coordinates from one normal orthogonal basis to another, since if $P = \pm 1$, $|P^{-1}AP| = |P^{-1}| \, |A| \, |P| = |A|$.

DEFINITION. A proper orthogonal transformation of $\mathscr{V}_n(\mathscr{R})$ is called a *rotation*. This amounts to defining a rotation of $\mathscr{V}_n(\mathscr{R})$ as any linear transformation of the space that preserves lengths, angles, and orientations.

This concept of a rotation, although similar to the rotation of axes considered in Chapter 5, should not be confused with it. In Chapter 5, the basis vectors were rotated to a new position and the new coordinates of a fixed vector were given in terms of the old by means of a proper orthogonal matrix. Here each vector of the space is mapped onto an image vector in such a way that every vector may be brought into the position originally occupied by its image by rotating the entire space. The coordinates of the image vector are given in terms of the coordinates of its antecedent (both relative to the same basis!) by means of a proper orthogonal matrix.

To justify our general definition of a rotation, we shall now show that every proper orthogonal transformation of $\mathscr{V}_3(\mathscr{R})$ is a rotation in the ordinary sense of the word. We shall show how to find the axis of rotation and the angle through which a plane perpendicular to this axis is rotated.

Let A be the matrix of a proper orthogonal transformation τ of $\mathscr{V}_3(\mathscr{R})$ relative to the basis $E_1 = (1, 0, 0)$, $E_2 = (0, 1, 0)$, $E_3 = (0, 0, 1)$. Since $|A| = |A^t| = 1$, we may write

$$|A - I| = |A - I| \, |A^t| = |AA^t - A^t| = |I - A^t| = (-1)^3 |A^t - I| = -|A - I|.$$

Hence $|A - I| = 0$, and $A - I$ is a singular matrix. It follows from Theorem 2.4 that there exists a nonzero vector S such that $(A - I)^t S^t = O$ and therefore $S(A - I) = O$, or $SA = S$. Since SA is the image of S under the transformation τ, we see that there exists a nonzero vector S that is invariant under τ. Similarly every scalar multiple of S is invariant under τ. Now let $\mathscr{S}$ be the subspace of $\mathscr{V}_3(\mathscr{R})$ generated by S, and let $\mathscr{T}$ be the orthogonal complement of $\mathscr{S}$ in $\mathscr{V}_3(\mathscr{R})$, that is, the plane perpendicular to S. It is clear that $\mathscr{S}$ is invariant under τ. Moreover, since τ is orthogonal, it preserves angles, and therefore every vector orthogonal to $\mathscr{S}$ is mapped by τ onto another vector orthogonal to $\mathscr{S}$. Since $\mathscr{T}$ consists of all those vectors of $\mathscr{V}_3(\mathscr{R})$ that are orthogonal to $\mathscr{S}$, and, since by Theorem 6.6 the dimension of $\mathscr{T}$ is preserved, it follows that $\mathscr{T}$ is invariant under τ and is actually mapped *onto* itself.

Now let F_3 be a unit vector in the space $\mathscr{S}$, and let F_1 and F_2 be any normal orthogonal basis of $\mathscr{T}$. Hence F_1, F_2, F_3 constitute a normal orthogonal basis of $\mathscr{V}_3(\mathscr{R})$. Since F_3 is invariant under τ and since every vector of $\mathscr{T}$ is mapped by τ on a vector of $\mathscr{T}$, the images of the basis vectors F_1, F_2, F_3 are given by equations of the form

$$F_1' = aF_1 + bF_2,$$
$$F_2' = cF_1 + dF_2,$$
$$F_3' = F_3.$$

Hence the matrix of τ relative to the F-basis is

$$P^{-1}AP = \begin{pmatrix} a & b & 0 \\ c & d & 0 \\ 0 & 0 & 1 \end{pmatrix},$$

where P is the matrix of the transformation of coordinates from the E-basis to the F-basis. Since both bases are normal orthogonal, P is orthogonal, and so, therefore, is $P^{-1}AP$. In fact, $P^{-1}AP$ is proper orthogonal since A is.

From the orthogonality of $P^{-1}AP$ we have $a^2 + b^2 = 1$, and so we may put $a = \cos\theta$ and $b = \pm\sin\theta$, where $0 \leq \theta \leq \pi$. Since also $a^2 + c^2 = 1$, we have $c = \pm\sin\theta$ and, from $c^2 + d^2 = 1$, $d = \pm\cos\theta$. The possibilities for the matrix

$$\begin{pmatrix} a & b \\ c & d \end{pmatrix}$$

are therefore

$$\begin{pmatrix} \cos\theta & \pm\sin\theta \\ \pm\sin\theta & \pm\cos\theta \end{pmatrix}$$

The further orthogonality conditions $ab + cd = 0$, $ac + bd = 0$ together with the requirement that $|P^{-1}AP| = ad - bc = 1$ eliminate all combinations of signs except

$$\begin{pmatrix} \cos \theta & -\sin \theta \\ \sin \theta & \cos \theta \end{pmatrix} \text{ and } \begin{pmatrix} \cos \theta & \sin \theta \\ -\sin \theta & \cos \theta \end{pmatrix}$$

The matrix of the linear transformation induced in $\mathscr{T}$ by the transformation τ therefore has one of these two forms. Comparison with Example 4, Section 33, shows that the first represents a rotation of the plane $\mathscr{T}$ through an angle $-\theta$ and the second a rotation through an angle θ. By Theorem 4.2 every vector X of $\mathscr{V}_3(\mathscr{R})$ has the form $X = S + T$, where $S \in \mathscr{S}$ and is invariant under τ, whereas $T \in \mathscr{T}$ and is rotated by τ through a fixed angle. It is therefore clear that $X' = S + T'$ and that the effect of τ is to rotate the whole space $\mathscr{V}_3(\mathscr{R})$ about the line $\mathscr{S}$ as axis.

If B is any improper orthogonal matrix of order n, it may be converted into a proper orthogonal matrix by changing the signs of all elements in its first row. Hence we may write $B = QA$, where A is proper orthogonal and

$$Q = \begin{pmatrix} -1 & 0 & \cdots & 0 \\ 0 & 1 & \cdots & 0 \\ \cdot & \cdot & \cdot & \cdot \\ 0 & 0 & \cdots & 1 \end{pmatrix}$$

The linear transformation whose matrix relative to a normal orthogonal basis $E_1, \cdots, E_n$ is Q is a reflection of the space $\mathscr{V}_n(\mathscr{R})$ in the hyperplane generated by $E_2, E_3, \cdots, E_n$. Hence every improper orthogonal transformation of $\mathscr{V}_n(\mathscr{R})$ is equivalent to a reflection in a coordinate hyperplane followed by a rotation of the space.

Exercise 6.4

1. Show that the linear transformation of $\mathscr{V}_3(\mathscr{R})$ whose matrix relative to the basis $(1, 0, 0)$, $(0, 1, 0)$, $(0, 0, 1)$ is

$$\begin{pmatrix} \tfrac{2}{3} & \tfrac{2}{3} & \tfrac{1}{3} \\ -\tfrac{2}{3} & \tfrac{1}{3} & \tfrac{2}{3} \\ \tfrac{1}{3} & -\tfrac{2}{3} & \tfrac{2}{3} \end{pmatrix}.$$

is a rotation. Find the axis of rotation and the angle through which the plane perpendicular to this axis is rotated.

2. Prove that a rotation in a space of an odd number of dimensions always leaves a one-dimensional subspace invariant.

Linear Transformations

3. Find the matrix relative to the basis $F_1 = (\tfrac{2}{3}, -\tfrac{2}{3}, \tfrac{1}{3})$, $F_2 = (\tfrac{2}{3}, \tfrac{1}{3}, -\tfrac{2}{3})$, $F_3 = (\tfrac{1}{3}, \tfrac{2}{3}, \tfrac{2}{3})$ of (a) a counterclockwise rotation through $45°$ about the x-axis; (b) a reflection in the xy-plane.

4. Find the matrix relative to the basis $E_1 = (1, 0, 0)$, $E_2 = (0, 1, 0)$, $E_3 = (0, 0, 1)$ of (a) a reflection in the yz-plane; (b) a reflection in the plane $x - y = 0$; (c) a reflection in the plane $2x + y - z = 0$; (d) a counterclockwise rotation through $45°$ about the x-axis; (e) a counterclockwise rotation through $120°$ about the line $x = y = z$.

5. If A is an improper orthogonal matrix, prove that $|A + I| = 0$.

6. Prove that the product of two reflections is a rotation.

7. If τ is the reflection in $\mathscr{V}_3(\mathscr{R})$ described in Problem 4 (b) and σ is the reflection in Problem 4 (c), find the matrix of the product transformation $\tau\sigma$, show that $\tau\sigma$ is a rotation, and find the axis about which this rotation takes place.

8. Let σ be a reflection in a plane p_1 and τ be a reflection in a plane p_2. Show that the transformation $\sigma\tau$ leaves the line of intersection of p_1 and p_2 invariant and must therefore represent a rotation about this line.

37. INVARIANT SUBSPACES AND THE REDUCTION OF THE MATRIX OF A LINEAR TRANSFORMATION

The investigation in the preceding section of a general rotation in $\mathscr{V}_3(\mathscr{R})$ by finding subspaces $\mathscr{S}$ and $\mathscr{T}$ that are left invariant illustrates a general method of considerable importance. Let $\mathscr{S}$ be any space of dimension n, and let $\mathscr{T}$ be any subspace of $\mathscr{S}$ of dimension $r < n$. Suppose τ is a linear transformation of $\mathscr{S}$ that leaves $\mathscr{T}$ invariant. By Theorem 1.9 we can choose a basis $E_1, E_2, \cdots, E_r, \cdots, E_n$ of $\mathscr{S}$ such that the first r vectors $E_1, \cdots, E_r$ constitute a basis of $\mathscr{T}$. Now let $E_1', E_2', \cdots, E_n'$ be the images under τ of these basis vectors. Since $\mathscr{T}$ is invariant, $E_1', E_2', \cdots, E_r'$ belong to $\mathscr{T}$, and hence the equations defining the E_i' in terms of the E_i have the form

$$E_1' = a_{11}E_1 + \cdots + a_{1r}E_r,$$
$$\cdots \cdots \cdots \cdots \cdots$$
$$E_r' = a_{r1}E_1 + \cdots + a_{rr}E_r,$$
$$E_{r+1}' = a_{r+1,1}E_1 + \cdots + a_{r+1,r}E_r + \cdots + a_{r+1,n}E_n,$$
$$\cdots \cdots \cdots \cdots \cdots$$
$$E_n' = a_{n1}E_1 + \cdots + a_{nr}E_r + \cdots + a_{nn}E_n,$$

and the matrix of τ relative to the E-basis is

$$
(5) \quad \begin{pmatrix} a_{11} & \cdots & a_{1r} & 0 & \cdots & 0 \\ \vdots & & \vdots & \vdots & & \vdots \\ a_{r1} & \cdots & a_{rr} & 0 & \cdots & 0 \\ a_{r+1\,1} & \cdots & a_{r+1\,r} & & \cdots & a_{r+1\,n} \\ \vdots & & \vdots & & & \vdots \\ a_{n1} & \cdots & a_{nr} & & \cdots & a_{nn} \end{pmatrix}
$$

Such a transformation τ that leaves a proper subspace $\mathscr{T} \neq O$ invariant is said to be *reducible*, and (5) is the *reduced form* of its matrix. The matrix of order r in the upper left-hand corner of (5) is the matrix giving the images of the basis vectors $E_1, \cdots, E_r$ of $\mathscr{T}$ in terms of these basis vectors. It is therefore the matrix of the transformation induced by τ in the subspace $\mathscr{T}$.

Now it may happen that the subspace $\mathscr{U}$ of $\mathscr{S}$ which is generated by the vectors $E_{r+1}, E_{r+2}, \cdots, E_n$ is also left invariant by τ. If this is the case, then $E'_{r+1}, \cdots, E'_n$ are linear combinations of the last $n - r$ basis vectors $E_{r+1}, \cdots, E_n$, and hence all the coefficients $a_{r+1\,1}, \cdots, a_{r+1\,r}$, $\cdots, a_{n1}, \cdots, a_{nr}$ are zero. In this case the matrix of τ takes the form

$$
(6) \quad \begin{pmatrix} A_1 & O \\ O & B_1 \end{pmatrix},
$$

where A_1 and B_1 are square matrices of order r and $n - r$, respectively. The transformation τ is then said to be *completely reducible*, and (6) is the reduced form of its matrix. Reducibility of τ need not imply complete reducibility. See, for example, Exercise 6.5, Problem 2. If the spaces $\mathscr{T}$ and $\mathscr{U}$ in turn have proper subspaces that are invariant under τ, then this process of reduction can be carried further. In order to state these results in their most convenient form, we must introduce some additional terms.

DEFINITION. A space $\mathscr{S}$ is said to be the *direct sum* of subspaces $\mathscr{T}_1, \mathscr{T}_2, \cdots, \mathscr{T}_m$ if (a) $\mathscr{S}$ is the sum of these subspaces and (b) the intersection of each subspace $\mathscr{T}_i$ with the sum of the $m - 1$ remaining subspaces is O.

We denote the fact that $\mathscr{S}$ is the direct sum of $\mathscr{T}_1, \mathscr{T}_2, \cdots, \mathscr{T}_m$ by writing

$$\mathscr{S} = \mathscr{T}_1 \dotplus \mathscr{T}_2 \dotplus \cdots \dotplus \mathscr{T}_m.$$

It follows from the definition that, if $\mathscr{S} = \mathscr{T}_1 \dotplus \mathscr{U}$ and $\mathscr{U} = \mathscr{T}_2 \dotplus \mathscr{T}_3 \dotplus \cdots \dotplus \mathscr{T}_m$, then $\mathscr{S} = \mathscr{T}_1 \dotplus \mathscr{T}_2 \dotplus \cdots \dotplus \mathscr{T}_m$. Thus the operation $\dotplus$ is associative. It is also clearly commutative since the order in which the subspaces are written is immaterial.

Theorem 6.11. If $\mathscr{S} = \mathscr{T}_1 \dotplus \mathscr{T}_2 \dotplus \cdots \dotplus \mathscr{T}_m$, every vector S in $\mathscr{S}$ has a unique representation in the form

$$S = T_1 + T_2 + \cdots + T_m,$$

where $T_i \in \mathscr{T}_i$ ($i = 1, 2, \cdots, m$).

Proof. Since $\mathscr{S}$ is the sum of $\mathscr{T}_1, \mathscr{T}_2, \cdots, \mathscr{T}_m$, S certainly has such a representation. If there were two such, we would have

$$S = T_1 + T_2 + \cdots + T_m = T_1' + T_2' + \cdots + T_m',$$

where $T_i' \in \mathscr{T}_i$, and therefore

$$T_1 - T_1' = (T_2' - T_2) + \cdots + (T_m' - T_m).$$

It follows, since $T_i' - T_i \in \mathscr{T}_i$, that $T_1 - T_1'$ belongs both to $\mathscr{T}_1$ and to $\mathscr{T}_2 + \mathscr{T}_3 + \cdots + \mathscr{T}_m$. Hence by the definition of direct sum, $T_1 - T_1' = O$ and $T_1 = T_1'$. Similarly $T_i' = T_i$ for $i = 2, 3, \cdots, m$.

Theorem 6.12. If $\mathscr{S} = \mathscr{T}_1 \dotplus \mathscr{T}_2 \dotplus \cdots \dotplus \mathscr{T}_m$, where $\mathscr{T}_i$ has dimension r_i and $\mathscr{S}$ has dimension n, then $n = r_1 + r_2 + \cdots + r_n$ and there exists a basis of $\mathscr{S}$ of which the first r_1 vectors constitute a basis of $\mathscr{T}_1$, the next r_2 vectors a basis of $\mathscr{T}_2, \cdots$, and the last r_m vectors a basis of $\mathscr{T}_m$.

Proof. The theorem is an immediate consequence of Theorems 1.9 and 1.10 and will be left as an exercise for the student.

Theorem 6.13. Let $\mathscr{S}$ be a space of dimension n that is the direct sum of subspaces $\mathscr{T}_i$ of dimension r_i ($i = 1, 2, \cdots, m$). If τ is a linear transformation of $\mathscr{S}$ that leaves each of the subspaces $\mathscr{T}_i$ invariant, then τ is completely reducible and its matrix relative to a suitably chosen basis of $\mathscr{S}$ has the form

(7)
$$A = \begin{pmatrix} A_1 & O & \cdots & O \\ O & A_2 & \cdots & O \\ \cdot & \cdot & \cdot & \cdot \\ O & O & \cdots & A_m \end{pmatrix}$$

where A_i is a square matrix of order r_i ($i = 1, 2, \cdots, m$) and all elements of A not in the diagonal blocks $A_1, \cdots, A_m$ are zero.

Proof. Choose a basis $E_1, \cdots, E_{r_1}, E_{r_1+1}, \cdots, E_{r_1+r_2}, \cdots, E_n$ as in Theorem 6.12, so that $E_1, \cdots, E_{r_1}$ is a basis of $\mathcal{T}_1$, $E_{r_1+1}, \cdots, E_{r_1+r_2}$ is a basis of $\mathcal{T}_2$, etc. Since τ leaves each of these subspaces invariant, the images $E_1', \cdots, E_{r_1}'$ are linear combinations of $E_1, \cdots, E_{r_1}$ only. Similarly $E_{r_1+1}', \cdots, E_{r_1+r_2}'$ are linear combinations of $E_{r_1+1}, \cdots, E_{r_1+r_2}$, etc. It follows that the matrix A of τ relative to this basis has the form (7). Moreover each matrix A_i is the matrix of the linear transformation induced in $\mathcal{T}_i$ by τ since the images of the basis vectors of $\mathcal{T}_i$ are given in terms of these basis vectors by the matrix A_i.

Note that, if τ is a magnification in $\mathcal{S}$, there are n distinct one-dimensional subspaces of $\mathcal{S}$ that are left invariant by τ, and $\mathcal{S}$ is the direct sum of these subspaces. Hence the matrix of τ takes the form (7), where each A_i is a matrix of order 1, and A is therefore a diagonal matrix. Another example of the operation of Theorem 6.13 is provided by the reduction of the matrix of a rotation in $\mathcal{V}_2(\mathcal{R})$ carried through in the last section. Later we shall investigate the general rotation of $\mathcal{V}_n(\mathcal{R})$ by use of Theorem 6.13.

Exercise 6.5

1. Write out the proof of Theorem 6.12 in detail.

2. Prove that the shear in $\mathcal{V}_2(\mathcal{R})$ defined by the equations

$$x' = x + ky,$$
$$y' = y \qquad k \neq 0$$

is not completely reducible.

3. Show that a rotation in $\mathcal{V}_2(\mathcal{R})$, through an angle that is not a multiple of $180°$, is never reducible but that a rotation in $\mathcal{V}_3(\mathcal{R})$ is always completely reducible.

4. Explain how to choose a basis of $\mathcal{V}_3(\mathcal{R})$ so that the corresponding matrix of a given rotation has the reduced form

$$\begin{pmatrix} 1 & 0 \\ 0 & A \end{pmatrix}$$

where A is a 2nd-order matrix. What is the form of A?

5. Explain how to choose a basis of $\mathcal{V}_n(\mathcal{R})$, where n is any odd number greater than 1, so that a given rotation of $\mathcal{V}_n(\mathcal{R})$ has corresponding matrix of the form

$$\begin{pmatrix} 1 & 0 \\ 0 & A \end{pmatrix}$$

where A is an $(n-1)$th-order matrix. (See Problem 2, Exercise 6.4.)

6. If a space $\mathscr{S}$ is the direct sum of subspaces $\mathscr{T}_1, \cdots, \mathscr{T}_m$, show that

(a) The set of all nonsingular linear transformations of $\mathscr{S}$ that leave the subspaces $T_1, \cdots, T_m$ invariant is a group $\mathscr{G}$.

(b) If τ and σ are two transformations in $\mathscr{G}$ whose matrices relative to a given basis are A_τ and A_σ, then $A_\tau A_\sigma = A_{\tau\sigma}$ and hence the matrices $A_\tau, \tau \in \mathscr{G}$ also form a group.

(c) There exists a matrix P, independent of τ, such that, for all τ in $\mathscr{G}$, $P^{-1}A_\tau P$ has the form (7).

CHAPTER SEVEN

Similarity of Matrices and Diagonalization Theorems

38. THE CHARACTERISTIC ROOTS AND EIGENVECTORS OF A MATRIX

Let τ be any linear transformation of $\mathscr{V}_n(\mathscr{R})$, and let A be its matrix relative to the basis $E_1 = (1, 0, \cdots, 0)$, $E_2 = (0, 1, 0, \cdots, 0), \cdots$, $E_n = (0, 0, \cdots, 1)$. As we have suggested in the preceding chapter, considerable information concerning the geometric effect of τ can be deduced from algebraic properties of the matrix A. As an illustration of this, we shall determine the one-dimensional subspaces of $\mathscr{V}_n(\mathscr{R})$ that are left invariant by τ. It is clear that a one-dimensional subspace is left invariant by τ if and only if it is generated by a vector X that is mapped by τ onto a scalar multiple of itself. We therefore seek all nonzero vectors X of $\mathscr{V}_n(\mathscr{R})$ such that $XA = kX$, where k is a real number. This implies that $A^t X^t = k X^t$ which, in turn, may be written in the form

(1) $$(A^t - kI)X^t = O.$$

If $X = (x_1, x_2, \cdots, x_n)$, equation 1 is equivalent to the set of n homogeneous linear equations

(2) $$\begin{aligned}(a_{11} - k)x_1 + a_{21}x_2 + \cdots + a_{n1}x_n &= 0, \\ a_{12}x_1 + (a_{22} - k)x_2 + \cdots + a_{n2}x_n &= 0, \\ &\cdots \\ a_{1n}x_1 + a_{2n}x_2 + \cdots + (a_{nn} - k)x_n &= 0,\end{aligned}$$

whose matrix is $A^t - kI$. By Theorem 2.4, a nonzero solution vector of

Similarity and Diagonalization Theorems

these equations exists if and only if $A^t - kI$ is a singular matrix. We must therefore choose k to be any real solution of the equation

(3) $$|A^t - kI| = 0.$$

For every such k a real nonzero solution vector X of (1) can be found, and every such vector will generate a one-dimensional subspace of $\mathscr{V}_n(\mathscr{R})$ that is invariant under τ.

The expansion of the determinant $|A^t - kI|$, which is the left-hand side of (3), is a polynomial of degree n in k with real coefficients. Equation 3 therefore has n solutions for k in the complex field, although these may not be distinct. It is easy to see that only the real solutions for k can lead to real nonzero solution vectors of (2). For, if $k = c + id$, the existence of a real solution of the first equation would imply, on equating the imaginary part of the left-hand side to zero, that $x_1 d = 0$. Similarly, the remaining equations yield $x_2 d = 0, \cdots, x_n d = 0$, so that either $d = 0$, and k is real, or $x_1 = x_2 = \cdots = x_n = 0$.

In practice it is useful to consider the complex roots of (3) as well. These will yield, when substituted in (2), complex solution vectors X such that $XA = kX$. Thus, if we think of A as the matrix of a linear transformation in $\mathscr{V}_n(\mathscr{C})$, all nonzero solution vectors of (2), where k is any solution of (3), will generate invariant one-dimensional subspaces of $\mathscr{V}_n(\mathscr{C})$.

These results will be summarized in Theorem 7.1, but first we introduce several definitions that will simplify their statement.

DEFINITION. If A is any square matrix of order n with elements in the complex field, the determinant $|A - xI|$ is called the *characteristic determinant* of A. The expansion of this determinant is a polynomial of degree n in x called the *characteristic polynomial* of A. The equation $|A - xI| = 0$ is called the *characteristic equation* of A, and the n roots of this equation are called the *characteristic roots* of A.

The characteristic roots of a matrix are also called by some writers the *latent roots* of A or *eigenvalues* of A.

DEFINITION. If A is any nth-order matrix and k is any characteristic root of A, a nonzero vector X of $\mathscr{V}_n(\mathscr{C})$ that satisfies the equation $(A - kI)X^t = O$ is called an *eigenvector* of A.

Note that, since $A - kI$ is singular, an eigenvector of A always exists corresponding to each characteristic root k. Also any nonzero scalar multiple of an eigenvector is an eigenvector corresponding to the same characteristic root. We have therefore proved

THEOREM 7.1. (a) Let A be the matrix relative to the basis $(1, 0, \cdots, 0)$, $(0, 1, \cdots, 0), \cdots, (0, 0, \cdots, 1)$ of a linear transformation τ of $\mathscr{V}_n(\mathscr{C})$. Corresponding to each characteristic root of A^t there exists an eigenvector of A^t, and every eigenvector of A^t generates a one-dimensional subspace of $\mathscr{V}_n(\mathscr{C})$ that is left invariant by τ.

(b) Let A be the matrix (relative to the above basis) of a linear transformation τ of $\mathscr{V}_n(\mathscr{R})$. Corresponding to every real characteristic root of A^t, and only to these, there exists a real eigenvector of A^t, and every real eigenvector of A^t generates a one-dimensional subspace of $\mathscr{V}_n(\mathscr{R})$ that is left invariant by τ.

If τ is a singular transformation, then A^t is singular and will have a characteristic root equal to zero. Any corresponding eigenvector will generate an invariant subspace that is mapped by τ onto the zero vector. Thus, if τ is singular with matrix A, the linearly independent eigenvectors of A^t corresponding to the characteristic root zero will generate the kernel of τ. In fact the kernel consists precisely of all eigenvectors corresponding to zero together with the zero vector.

39. SIMILARITY

We have seen in Section 35 that, if A and B are the matrices of the same linear transformation relative to two different bases of the space, then there exists a nonsingular matrix P such that $B = P^{-1}AP$. We shall now derive some properties of matrices A and B that are related in this way.

DEFINITION. A matrix B is said to be *similar* to a matrix A if there exists a nonsingular matrix P such that $B = P^{-1}AP$. The passage from A to $P^{-1}AP$ is called a *similarity transformation*, and $P^{-1}AP$ is called the *transform* of A by P.

THEOREM 7.2. Similarity of matrices is an equivalence relation.

Proof. Using $P = I$, we see that A is similar to itself, and similarity is therefore reflexive. If $B = P^{-1}AP$, then $A = PBP^{-1} = Q^{-1}BQ$, where $Q = P^{-1}$. Hence similarity is symmetric. Finally, if $B = P^{-1}AP$ and $C = R^{-1}BR$, then $C = R^{-1}P^{-1}APR = S^{-1}AS$, where $S = PR$. This proves the transitivity.

In view of the symmetry of the similarity relationship we may speak without ambiguity of "similar matrices A and B."

THEOREM 7.3. Similar matrices have equal determinants, the same characteristic equations, and the same characteristic roots.

Similarity and Diagonalization Theorems

Proof. If $B = P^{-1}AP$, then $|B| = |P^{-1}| |A| |P| = |P^{-1}| |P| |A| = |P^{-1}P| |A| = |A|$. Hence A and B have equal determinants. It then follows that $|A - xI| = |P^{-1}(A - xI)P| = |P^{-1}AP - xP^{-1}IP| = |P^{-1}AP - xI|$. Thus A and $P^{-1}AP$ have the same characteristic polynomials, and hence the same characteristic equations and the same characteristic roots, but not necessarily the same eigenvectors. (See Problem 3, Exercise 7.1.)

In view of Theorem 7.3 we may now speak of the characteristic roots of a linear transformation τ of a space $\mathscr{S}$, meaning thereby the characteristic roots of the matrix of τ relative to any basis. For Theorems 6.8 and 7.3 tell us that these characteristic roots are independent of the basis chosen.

Exercise 7.1

1. Find the characteristic roots and eigenvectors of the following matrices:

    (a) $\begin{pmatrix} 2 & 4 \\ 3 & 13 \end{pmatrix}$ (b) $\begin{pmatrix} 2 & -3 \\ -3 & 1 \end{pmatrix}$ (c) $\begin{pmatrix} 3 & -2 \\ 2 & 1 \end{pmatrix}$

    (d) $\begin{pmatrix} 3 & 2 & 4 \\ 2 & 0 & 2 \\ 4 & 2 & 3 \end{pmatrix}$ (e) $\begin{pmatrix} \frac{2}{3} & \frac{2}{3} & \frac{1}{3} \\ -\frac{2}{3} & \frac{1}{3} & \frac{2}{3} \\ \frac{1}{3} & -\frac{2}{3} & \frac{2}{3} \end{pmatrix}$

2. What are the characteristic roots and eigenvectors of the unit matrix?

3. If X is an eigenvector of a matrix A^t, prove that XP is an eigenvector of $(P^{-1}AP)^t$ corresponding to the same characteristic root. Prove this result both algebraically and by interpreting it in terms of invariant subspaces under the linear transformation whose matrix is A.

4. Prove that a linear transformation τ of $\mathscr{V}_n(\mathscr{R})$ or $\mathscr{V}_n(\mathscr{C})$ leaves a nonzero vector invariant if and only if the matrix of τ has a characteristic root equal to one.

5. Prove that every proper orthogonal transformation in a space of an odd number of dimensions leaves a nonzero vector invariant. *Hint:* proof for $\mathscr{V}_3(\mathscr{R})$ is given in Section 36.

6. Show that for every rotation of axes in three-dimensional space (see Section 30, Chapter 5) there is a line through the origin every point of which has the same coordinates relative to the new axes as it had relative to the old axes. How can the direction of this "invariant line" be determined?

7. Find the direction of the invariant line for the rotation from the basis vectors $(1, 0, 0)$, $(0, 1, 0)$, $(0, 0, 1)$ to the basis vectors $(\frac{2}{3}, \frac{2}{3}, \frac{1}{3})$, $(-\frac{2}{3}, \frac{1}{3}, \frac{2}{3})$, $(\frac{1}{3}, -\frac{2}{3}, \frac{2}{3})$.

8. If $A = \begin{pmatrix} a & b \\ c & d \end{pmatrix}$ and $Q = \begin{pmatrix} 0 & 1 \\ 1 & 0 \end{pmatrix}$, find $Q^{-1}PQ$.

9. If A is any matrix of order n, the *secondary diagonal* of A is the diagonal from the lower left-hand to the upper right-hand corner of A. Let Q be the matrix of order n with all its secondary diagonal elements equal to one and all other elements equal to zero. Show that the effect of transforming A by Q is to rotate the elements of A about its central point through an angle of $180°$; that is, to shift the element in the ith row and jth column of A to the $(n - j + 1)$th row and $(n - i + 1)$th column, $(i, j = 1, 2, \cdots, n)$. Verify this by actual multiplication for the general third-order matrix $A = (a_{ij})$.

10. Show that transforming A by the matrix Q defined in Problem 9 has the effect of reflecting A first in its secondary diagonal and then in its main diagonal.

11. Deduce from Problem 10, that, if A is symmetric about its secondary diagonal, then $Q^{-1}AQ = A^t$.

12. If A^s is the matrix obtained by reflecting A in its secondary diagonal, show that $Q^{-1}AQ = (A^s)^t = (A^t)^s$, and hence that $|A^s| = |A|$.

40. MATRICES THAT ARE SIMILAR TO DIAGONAL MATRICES

Most of the rest of this chapter will be devoted to the problem of the reduction of a given matrix to diagonal form by means of a similarity transformation. This reduction is not always possible, and a complete discussion of the problem will not be given. Certain special cases, however, will be considered in detail.

Suppose first that a given nth-order matrix A is similar to a diagonal matrix D so that there exists a nonsingular matrix S such that $S^{-1}AS = D$. By Theorem 7.3, A and D have the same characteristic roots. But the characteristic roots of a diagonal matrix are clearly the elements in the main diagonal. Hence the diagonal elements of D are precisely the characteristic roots of A. We therefore have

THEOREM 7.4. If a matrix A is similar to a diagonal matrix D, then

$$D = \begin{pmatrix} k_1 & 0 & \cdots & 0 \\ 0 & k_2 & \cdots & 0 \\ \cdot & \cdot & \cdot & \cdot \\ 0 & 0 & \cdots & k_n \end{pmatrix}$$

where $k_1, k_2, \cdots, k_n$ are the characteristic roots of A.

Again, suppose that A is similar to a diagonal matrix D so that $S^{-1}AS = D$ or $AS = SD$. Let $A = (a_{ij})$, $S = (s_{ij})$, and let $k_1, k_2, \cdots, k_n$

Similarity and Diagonalization Theorems

be the characteristic roots of A, and therefore the diagonal elements of D. We then have

$$AS = SD = \begin{pmatrix} s_{11} & s_{12} & \cdots & s_{1n} \\ s_{21} & s_{22} & \cdots & s_{2n} \\ \cdot & \cdot & \cdot & \cdot \\ s_{n1} & s_{n2} & \cdots & s_{nn} \end{pmatrix} \begin{pmatrix} k_1 & 0 & \cdots & 0 \\ 0 & k_2 & \cdots & 0 \\ \cdot & \cdot & \cdot & \cdot \\ 0 & 0 & \cdots & k_n \end{pmatrix}$$

$$= \begin{pmatrix} k_1 s_{11} & k_2 s_{12} & \cdots & k_n s_{1n} \\ k_1 s_{21} & k_2 s_{22} & \cdots & k_n s_{2n} \\ \cdot & \cdot & \cdot & \cdot \\ k_1 s_{n1} & k_2 s_{n2} & \cdots & k_n s_{nn} \end{pmatrix}$$

Now, if $S_1^t, S_2^t, \cdots, S_n^t$ are the column vectors of S, we see that the column vectors of SD are $k_1 S_1^t, k_2 S_2^t, \cdots, k_n S_n^t$, and (by Problem 11, Exercise 3.2) those of AS are $AS_1^t, AS_2^t, \cdots, AS_n^t$. Equating these, we get

(4) $\qquad AS_i^t = k_i S_i^t \qquad (i = 1, 2, \cdots, n).$

Equation 4 states that the ith column vector of S is an eigenvector of A corresponding to the characteristic root k_i. Conversely, any matrix S whose ith column vector is an eigenvector of A corresponding to the root k_i $(i = 1, 2, \cdots, n)$ will satisfy the matrix equation $AS = SD$. If, in addition, S is nonsingular then $S^{-1}AS = D$ and A is similar to a diagonal matrix. This completes the proof of

THEOREM 7.5. A necessary and sufficient condition that an nth-order matrix A, with complex elements, be similar to a diagonal matrix is that A have n linearly independent eigenvectors in $\mathscr{V}_n(\mathscr{R})$. Moreover $S^{-1}AS$ is a diagonal matrix if and only if the column vectors of S are linearly independent eigenvectors of A.

It should be observed that, if A is a real matrix, the diagonal form and the transforming matrix S (when they exist) will not be real unless all the characteristic roots of A are real. Only in this case, therefore, are our results capable of immediate interpretation in terms of linear transformations of $\mathscr{V}_n(\mathscr{R})$.

THEOREM 7.6. If the characteristic roots of a matrix A are distinct, then A is similar to a diagonal matrix.

Proof. In view of Theorem 7.5, we need only prove that A has n linearly independent eigenvectors. Let $k_1, k_2, \cdots, k_n$ be the characteristic roots of A, and let S_i be an eigenvector of A corresponding to k_i ($i = 1, 2, \cdots, n$). Suppose that $S_1, S_2, \cdots, S_n$ are linearly dependent. We can then choose r so that $1 \leq r < n$ and $S_1, S_2, \cdots, S_{r+1}$ are linearly dependent while $S_1, S_2, \cdots, S_r$ are linearly independent. Let

(5) $$c_1 S_1^t + c_2 S_2^t + \cdots + c_{r+1} S_{r+1}^t = O$$

be a nontrivial relation among $S_1^t, \cdots, S_{r+1}^t$. Multiplying equation 5 on the left by A, and using $AS_i^t = k_i S_i^t$, we get

(6) $$c_1 k_1 S_1^t + c_2 k_2 S_2^t + \cdots + c_{r+1} k_{r+1} S_{r+1}^t = O.$$

Now, multiplying (5) by k_{r+1} and subtracting from (6), we find

$$c_1(k_1 - k_{r+1}) S_1^t + c_2(k_2 - k_{r+1}) S_2^t + \cdots + c_r(k_r - k_{r+1}) S_r^t = O.$$

Since $S_1^t, \cdots, S_r^t$ are linearly independent and $k_i \neq k_{r+1}$ for $i = 1, 2, \cdots, r$, the last equation implies $c_1 = c_2 = \cdots = c_r = 0$, and, since $S_{r+1} \neq O$ (5) then implies $c_{r+1} = 0$, in contradiction to the original assumption that (5) was a nontrivial relation. This contradiction shows that $S_1, S_2, \cdots, S_n$ must be linearly independent, and it follows from Theorem 7.5 that A is similar to a diagonal matrix.

If the characteristic roots of a matrix are not distinct, the matrix may not be similar to a diagonal matrix. Every matrix is, however, similar to a *triangular* matrix, that is, a matrix with only zeros below (or above) the main diagonal.

THEOREM 7.7. Let A be an nth-order matrix with complex elements and characteristic roots $k_1, k_2, \cdots, k_n$. There exists a nonsingular matrix P, with complex elements, such that

$$P^{-1}AP = \begin{pmatrix} k_1 & b_{12} & \cdots & b_{1n} \\ 0 & k_2 & \cdots & b_{2n} \\ \cdot & \cdot & \cdot & \cdot \\ 0 & 0 & \cdots & k_n \end{pmatrix}$$

Proof. Let S_1 be an eigenvector of A corresponding to k_1. Let S be any nonsingular matrix of which S_1^t is the first column vector. By Problem 11, Exercise 3.2, the first column vector of AS is $AS_1^t = k_1 S_1^t$, and the first column of $S^{-1}AS$ is therefore $S^{-1}(k_1 S_1^t)$, which is also the first column vector of $k_1(S^{-1}S)$, namely,

$$(k_1, 0, \cdots, 0)^t.$$

Hence we have
$$S^{-1}AS = \begin{pmatrix} k_1 & B_1 \\ O & A_1 \end{pmatrix}$$
where A_1 is a matrix of order $n-1$. Since
$$|S^{-1}AS - xI| = (k_1 - x)|A_1 - xI|,$$
and since, by Theorem 7.3, $S^{-1}AS$ and A have the same characteristic roots, it follows that the characteristic roots of A_1 are $k_2, k_3, \cdots, k_n$. The theorem is therefore proved for the case $n = 2$. We assume it true for matrices of order $n-1$ and proceed by induction. Since A_1 is of order $n-1$, there exists, by our induction assumption, a nonsingular matrix Q such that
$$Q^{-1}A_1Q = \begin{pmatrix} k_2 & b_{23} & \cdots & b_{2n} \\ 0 & k_3 & \cdots & b_{3n} \\ \cdot & \cdot & \cdot & \cdot \\ 0 & 0 & \cdots & k_n \end{pmatrix}$$

Now let
$$R = \begin{pmatrix} 1 & O \\ O & Q \end{pmatrix}$$
and we have
$$(SR)^{-1}A(SR) = R^{-1}S^{-1}ASR = \begin{pmatrix} 1 & O \\ O & Q^{-1} \end{pmatrix} \begin{pmatrix} k_1 & B_1 \\ O & A_1 \end{pmatrix} \begin{pmatrix} 1 & O \\ O & Q \end{pmatrix}$$
$$= \begin{pmatrix} k_1 & B_1Q \\ O & Q^{-1}A_1Q \end{pmatrix}$$
$$= \begin{pmatrix} k_1 & b_{12} & \cdots & b_{1n} \\ 0 & k_2 & \cdots & b_{2n} \\ \cdot & \cdot & \cdot & \cdot \\ 0 & 0 & \cdots & k_n \end{pmatrix}$$

The matrix $P = SR$ therefore transforms A to triangular form as required.

We note that, in view of the result of Problem 12, Exercise 7.2, the triangular form may be written, if we prefer, with the zeros above the main diagonal instead of below.

If
$$f(x) = a_0 x^m + a_1 x^{m-1} + \cdots + a_m$$

is any polynomial in x and A is any square matrix, we denote by $f(A)$ the matrix

$$a_0 A^m + a_1 A^{m-1} + \cdots + a_m I.$$

Using this notation we state the following important corollary of Theorem 7.7.

COROLLARY 1. If $k_1, k_2, \cdots, k_n$ are the characteristic roots of A and $f(x)$ is any polynomial, the characteristic roots of $f(A)$ are $f(k_1), f(k_2), \cdots, f(k_n)$.

Proof. If B and C are triangular matrices with zeros below the diagonal and diagonal elements $b_1, \cdots, b_n$ and $c_1, \cdots, c_n$, respectively, it is easy to verify that BC and $B + C$ are triangular with zeros below the diagonal and with diagonal elements $b_1 c_1, \cdots, b_n c_n$ and $b_1 + c_1, \cdots, b_n + c_n$, respectively. It follows that, if $P^{-1}AP$ is triangular with diagonal elements $k_1, \cdots, k_n$, then $f(P^{-1}AP)$ is triangular with diagonal elements $f(k_1), \cdots, f(k_n)$. But, since $(P^{-1}AP)^r = P^{-1}A^r P$ for any integer r, it follows from the distributive law for matrix multiplication that $f(P^{-1}AP) = P^{-1}f(A)P$. Hence $P^{-1}f(A)P$ is the triangular form of $f(A)$, and its diagonal elements $f(k_1), \cdots, f(k_n)$ are the characteristic roots of $f(A)$.

COROLLARY 2. If A is any matrix that is similar to a diagonal matrix and $f(x)$ is the characteristic polynomial of A, then $f(A) = O$.

Proof. Let $P^{-1}AP = D$, where D is the diagonal form of A. The diagonal elements of D are the characteristic roots $k_1, k_2, \cdots, k_n$ of A. We see, as in the proof of Corollary 1, that $P^{-1}f(A)P = f(P^{-1}AP) = f(D)$ is the diagonal form of $f(A)$. But by Corollary 1 the characteristic roots of $f(A)$, and therefore the diagonal elements of $f(D)$, are $f(k_1), \cdots, f(k_n)$. All of these, however, are zero since $k_1, \cdots, k_n$ are the roots of $f(x) = 0$. Hence $f(D) = O$, and therefore $f(A) = Pf(D)P^{-1} = O$.

Corollary 2 is a special case of the important Cayley-Hamilton theorem which states that, if A is any square matrix with characteristic polynomial $f(x)$, then $f(A) = O$. For the proof of this more general result the student is referred to [2] or [14].

It can be shown that every matrix A with complex elements is similar to a triangular matrix of the form

$$J = \begin{pmatrix} F_1 & O & \cdots & O \\ O & F_2 & \cdots & O \\ \cdot & \cdot & \cdot & \cdot \\ O & O & \cdots & F_r \end{pmatrix}$$

Similarity and Diagonalization Theorems

where each F_i has the form

$$\begin{pmatrix} k_i & 1 & 0 & \cdots & 0 \\ 0 & k_i & 1 & \cdots & 0 \\ \cdot & \cdot & \cdot & \cdot & \cdot \\ & & & & 1 \\ 0 & 0 & 0 & \cdots & k_i \end{pmatrix}$$

and $k_1, \cdots, k_r$ are the characteristic roots of A but are not necessarily distinct. This is called the *classical* or *Jordan canonical form* of A, and two matrices are similar if and only if they have the same Jordan canonical form except possibly for the order in which the matrices F_i occur in the diagonal of J. The Jordan canonical form is a diagonal matrix if and only if each of the matrices F_i has order 1. For the proof of these results the student is referred to [2], [11], or [13].

Exercise 7.2

1. Prove that the matrix

$$A = \begin{pmatrix} 1 & 2 & -4 \\ 0 & -1 & 6 \\ 0 & -1 & 4 \end{pmatrix}$$

is not similar to a diagonal matrix. Find a matrix that will reduce A to triangular form by a similarity transformation.

2. Reduce the following matrices to diagonal form by means of a similarity transformation:

(a) $\begin{pmatrix} -\frac{2}{7} & \frac{12}{7} \\ -\frac{40}{7} & \frac{23}{7} \end{pmatrix}$ (b) $\begin{pmatrix} 4 & 2 & -2 \\ -5 & 3 & 2 \\ -2 & 4 & 1 \end{pmatrix}$

(c) $\begin{pmatrix} \frac{8}{7} & \frac{3}{7} & -\frac{3}{7} \\ \frac{3}{7} & -\frac{6}{7} & \frac{3}{7} \\ \frac{3}{7} & \frac{3}{7} & \frac{8}{7} \end{pmatrix}$ (d) $\begin{pmatrix} 1 & 0 & 0 & -5 \\ 0 & -1 & 0 & 6 \\ 0 & 0 & 2 & 0 \\ 0 & 0 & 0 & 2 \end{pmatrix}$

3. If a linear transformation τ of $\mathscr{V}_n(\mathscr{C})$ has matrix A relative to the basis $(1, 0, \cdots, 0)$, $(0, 1, \cdots, 0)$, $\cdots$, $(0, 0, \cdots, 1)$ and if A^t has n distinct characteristic roots and hence n linearly independent eigenvectors $E_1, E_2, \cdots, E_n$, prove that the matrix of τ relative to the basis $E_1, E_2, \cdots, E_n$ is the diagonal matrix similar to A.

4. Prove that a linear transformation of $\mathscr{V}_n(\mathscr{R})$ whose characteristic roots are distinct positive real numbers is a magnification.

5. Prove that, if all the characteristic roots of A are real, there exists an orthogonal matrix P such that $P^{-1}AP$ is a triangular matrix.

6. If A is nonsingular, prove that the characteristic roots of A^{-1} are the reciprocals of those of A.

7. A nonzero matrix A is said to be nilpotent if, for some positive integer r, $A^r = O$. Show by successive multiplication that a triangular matrix in which all the diagonal elements are zero is nilpotent.

8. Show that a nonzero matrix is nilpotent if and only if all its characteristic roots are equal to zero.

9. Prove that a nilpotent matrix cannot be similar to a diagonal matrix.

10. If A is a matrix of order n and I is the unit matrix of order n, show that the characteristic roots of
$$B = \begin{pmatrix} O & I \\ A & O \end{pmatrix}$$
are $\pm\sqrt{\lambda_1}, \cdots, \pm\sqrt{\lambda_n}$, where $\lambda_1, \cdots, \lambda_n$ are the characteristic roots of A. *Hint:* reduce the characteristic determinant of B to a determinant of order n by successive application of D6 and D7, Section 15.

11. Any matrix that is similar to a diagonal matrix is similar to its transpose.

12. Prove that every matrix A is similar to its transpose. *Hint:* note that each of the matrices F_i in the Jordon normal form J of A is symmetric about its secondary diagonal. Hence, by Problem 11, Exercise 7.1, $Q_i^{-1} F_i Q_i = F_i^t$, where Q_i is a matrix of the type defined in Problem 9, Exercise 7.1. Let
$$Q = \begin{pmatrix} Q_1 & 0 & \cdots & 0 \\ 0 & Q_2 & \cdots & 0 \\ \cdot & \cdot & \cdot & \cdot \\ 0 & 0 & \cdots & Q_r \end{pmatrix}$$
and show that $Q^{-1}JG = J^t$. Now show that J^t is similar to A^t and the result follows.

41. SYSTEMS OF LINEAR DIFFERENTIAL EQUATIONS

Let $y_1, y_2, \cdots, y_n$ be functions of an independent variable x that satisfy the system of linear differential equations

(7) $$\frac{dy_i}{dx} = \sum_{j=1}^{n} a_{ij} y_j + b_i(x), \qquad (i = 1, 2, \cdots, n),$$

where $b_i(x)$ are integrable functions of x. The coefficients a_{ij} are in general functions of x but we shall consider only the important special

Similarity and Diagonalization Theorems

case in which a_{ij} are constants. It was shown in Section 20 that equations 7 can be written in the form

$$\frac{dY^t}{dx} = AY^t + B^t, \tag{8}$$

where $A = (a_{ij})$, $Y = (y_1, \cdots, y_n)$, and $B = (b_1, \cdots, b_n)$. To solve this system, we make a change of dependent variables and let

$$y_i = \sum_{j=1}^{n} p_{ij} z_j,$$

or, in matrix notation, let $Y^t = PZ^t$, where $P = (p_{ij})$ is a matrix of order n with constant elements. We therefore have

$$\frac{dY^t}{dx} = P\frac{dZ^t}{dx},$$

and equation 8 becomes

$$P\frac{dZ^t}{dx} = APZ^t + B^t$$

or

$$\frac{dZ^t}{dx} = (P^{-1}AP)Z^t + P^{-1}B^t. \tag{9}$$

Now suppose that the matrix A has n linearly independent eigenvectors and therefore P can be chosen so that $P^{-1}AP$ is a diagonal matrix. If the characteristic roots of A are $k_1, k_2, \cdots, k_n$, equation 9 is then equivalent to

$$\frac{dz_i}{dx} = k_i z_i + h_i(x) \qquad (i = 1, 2, \cdots, n),$$

where $h_i(x)$ is a linear combination, with constant coefficients, of $b_1(x)$, $\cdots, b_n(x)$. These equations are immediately integrable, their solutions being

$$z_i = c_i e^{k_i x} + e^{k_i x} \int e^{-k_i x} h_i(x) \, dx.$$

The above solution of equations 7 is theoretically extremely simple. To apply it, however, to a numerical example it is necessary first to find the characteristic roots of A and then to find the linearly independent eigenvectors of A that constitute the column vectors of P. If n is not larger than 5, this can be done by direct computation of the characteristic equation $|A - xI| = 0$ and subsequent solution of it by some method of successive approximation. For large values of n, however, this is quite impractical. For shorter methods of finding the characteristic roots of a matrix and solving systems of linear equations, the student is referred to [5] and [8].

The solution of equations 7 when A is not similar to a diagonal matrix can be made to depend on Theorem 7.7. If we choose P so that $P^{-1}AP$ is triangular (with zeros *above* the main diagonal), equation 9 is equivalent to the system

(10)
$$\frac{dz_1}{dx} = k_1 z_1 + h_1(x),$$
$$\frac{dz_2}{dx} = b_{21} z_1 + k_2 z_2 + h_2(x),$$
$$\cdots \cdots \cdots \cdots$$
$$\frac{dz_n}{dx} = b_{n1} z_1 + b_{n2} z_2 + \cdots + k_n z_n + h_n(x).$$

The first equation of (10) is linear in z_1 and hence immediately solvable. Its solution can be substituted in the second equation, which then becomes a linear equation in z_2. In general the solutions of the first $r - 1$ equations for $z_1, z_2, \cdots, z_{r-1}$ can be substituted in the rth equation, which can then be solved for z_r. If the Jordan canonical form of A is known, the solution would be still simpler since each equation would involve at most two of the y_i's.

If the original system is homogeneous, we have $b_i(x) = h_i(x) = 0$ ($i = 1, 2, \cdots, n$). In this case we can see what form the solutions of (10) will take without writing them down explicitly. From the first equation

$$z_1 = c_1 e^{k_1 x},$$

and the second equation becomes

$$\frac{dz_2}{dx} = k_2 z_2 + b_{21} c_1 e^{k_1 x}$$

This is a linear first-order equation in z_2 whose solution has the form

$$z_2 = c_2 e^{k_2 x} + c_3 e^{k_1 x} \qquad \text{if } k_2 \neq k_1,$$
$$z_2 = (c_2 x + c_3) e^{k_1 x} \qquad \text{if } k_2 = k_1,$$

and in each case $c_2 = 0$ if $b_{21} = 0$. Continuing thus, if $k_1 = k_2 = k_3$ the third equation will yield a solution of the form

$$z_3 = (c_4 x^2 + c_5 x + c_6) e^{k_1 x},$$

and in general r equal characteristic roots of A give rise to a solution of the form $e^{kx}p(x)$, where $p(x)$ is a polynomial of degree at most $r-1$. We can therefore conclude that, if $k_1, \cdots, k_m$ are the distinct characteristic roots of A, the solutions of the homogeneous system

$$\text{(11)} \qquad \frac{dy_i}{dx} = \sum_{j=1}^{n} a_{ij} y_j$$

are linear combinations of $e^{k_1 x}, \cdots, e^{k_m x}$ with coefficients that are either constants or polynomials in x whose degrees are bounded by the maximum multiplicity with which any one of the characteristic roots occurs. In applications it is often important to know whether the y_i's tend to 0 as $x \to \infty$. Our result shows that a necessary and sufficient condition that all the solutions of (11) tend to 0 as $x \to \infty$ is that all the characteristic roots of A have negative real parts.

As an application of the methods described above, consider an electric circuit containing a resistance R_1, capacitance C_1, and inductance L_1, connected in series. If this circuit is coupled with a similar circuit designated by R_2, C_2, and L_2, and if the mutual inductance between the circuits is M, then the currents I_1 and I_2 in the two circuits satisfy the differential equations

$$L_1 \frac{d^2 I_1}{dt^2} + M \frac{d^2 I_2}{dt^2} + R_1 \frac{dI_1}{dt} + \frac{I_1}{C_1} = 0,$$

$$M \frac{d^2 I_1}{dt^2} + L_2 \frac{d^2 I_2}{dt^2} + R_2 \frac{dI_2}{dt} + \frac{I_2}{C_2} = 0.$$

If we let $\dfrac{dI_1}{dt} = y_3$ and $\dfrac{dI_2}{dt} = y_4$, these equations are equivalent to the system

$$\frac{dI_1}{dt} = y_3,$$

$$\frac{dI_2}{dt} = y_4,$$

$$L_1 \frac{dy_3}{dt} + M \frac{dy_4}{dt} = -\frac{1}{C_1} I_1 - R_1 y_3,$$

$$M \frac{dy_3}{dt} + L_2 \frac{dy_4}{dt} = -\frac{1}{C_2} I_2 - R_2 y_4.$$

Denoting by Y the column vector $(I_1, I_2, y_3, y_4)^t$, these equations can now be written

$$A \frac{dY}{dt} = BY,$$

where

$$A = \begin{pmatrix} 1 & 0 & 0 & 0 \\ 0 & 1 & 0 & 0 \\ 0 & 0 & L_1 & M \\ 0 & 0 & M & L_2 \end{pmatrix}, \quad B = \begin{pmatrix} 0 & 0 & 1 & 0 \\ 0 & 0 & 0 & 1 \\ -\frac{1}{C_1} & 0 & -R_1 & 0 \\ 0 & -\frac{1}{C_2} & 0 & -R_2 \end{pmatrix}$$

We let $D = L_1 L_2 - M^2$. If $D \neq 0$, A is nonsingular and

$$\frac{dY}{dt} = A^{-1} BY,$$

where

$$A^{-1}B = \begin{pmatrix} 0 & 0 & 1 & 0 \\ 0 & 0 & 0 & 1 \\ -\frac{L_2}{C_1 D} & \frac{M}{C_2 D} & -\frac{R_1 L_2}{D} & \frac{M R_2}{D} \\ \frac{M}{C_1 D} & -\frac{L_1}{C_2 D} & \frac{R_1 M}{D} & -\frac{L_1 R_2}{D} \end{pmatrix}$$

If we now assume that the resistances R_1 and R_2 are negligible and put $R_1 = R_2 = 0$, the matrix $A^{-1}B$ takes the form that was considered in Problem 10, Exercise 7.2. Its characteristic roots are therefore $\pm \sqrt{\lambda}$ and $\pm \sqrt{\lambda'}$, where λ and λ' are the roots of

$$\begin{vmatrix} -\frac{L_2}{C_1 D} - x & \frac{M}{C_2 D} \\ \frac{M}{C_1 D} & -\frac{L_1}{C_2 D} - x \end{vmatrix} = 0.$$

This reduces to the quadratic equation

$$x^2 + \frac{L_1 C_1 + L_2 C_2}{D C_1 C_2} x + \frac{1}{D C_1 C_2} = 0,$$

Similarity and Diagonalization Theorems

whose discriminant is always positive, being equal to

$$\frac{(L_1C_1 - L_2C_2)^2 + 4M^2C_1C_2}{D^2C_1^2C_2^2}.$$

It follows that λ and λ' are real and unequal. Moreover, since the coefficient of x and the constant term agree in sign with D, a necessary and sufficient condition that λ and λ' be both negative is that $D > 0$ or $L_1L_2 > M^2$. Assuming that this condition is satisfied, we let $\lambda = -\omega^2$ and $\lambda' = -\omega'^2$, and the characteristic roots of $A^{-1}B$ are the pure imaginary numbers $\pm i\omega$ and $\pm i\omega'$. Since these are distinct, $A^{-1}B$ is similar to a diagonal matrix, and the solutions for I_1 and I_2 are therefore linear combinations of $e^{\pm i\omega t}$ and $e^{\pm i\omega' t}$ or of $\sin \omega t$, $\cos \omega t$, $\sin \omega' t$, and $\cos \omega' t$. The condition $L_1L_2 > M^2$ therefore ensures periodic solutions. Two frequencies occur in the solutions, namely, $\omega/2\pi$ and $\omega'/2\pi$. It is clear that, if the original differential equations are to be satisfied, both I_1 and I_2 must contain terms with frequency $\omega/2\pi$ and also terms with frequency $\omega'/2\pi$. In other words, both frequencies occur in both circuits. Once the frequencies are known the solutions for I_1 and I_2 could be determined by actually finding the linear transformation that diagonalizes $A^{-1}B$. However, it is probably easier to substitute expressions of the form $a_1 \cos \omega t + b_1 \sin \omega t + c_1 \cos \omega' t + d_1 \sin \omega' t$ for I_1 and I_2 and determine the relations between the unknown constants so that the differential equations are satisfied. There will be two arbitrary constants in the general solution for I_1 and two in the solution for I_2.

For additional information on the use of matrix methods in the theory of differential equations and their solution the student is referred to [1] and [5].

Exercise 7.3

1. Solve the following systems of differential equations:

(a) $\dfrac{dy_1}{dx} = y_2,$

$\dfrac{dy_2}{dx} = y_3,$

$\dfrac{dy_3}{dx} = 3y_1 + y_2 - 3y_3.$

(b) $\dfrac{dy_1}{dx} = y_2,$

$\dfrac{dy_2}{dx} = y_3,$

$\dfrac{dy_3}{dx} = y_3 - 4y_2 + 4y_1.$

(c) $\dfrac{dy_1}{dx} = 3y_1 - y_2,$

$\dfrac{dy_2}{dx} = y_1 + 3y_2.$

(d) $\dfrac{dy_1}{dx} = y_1 + y_2 - y_3,$

$\dfrac{dy_2}{dx} = y_1 - y_2 + y_3,$

$\dfrac{dy_3}{dx} = -y_1 + y_2 + y_3.$

2. Using the result of Problem 1, Exercise 7.2, solve the system

$$\frac{dy_1}{dx} = y_1 + 2y_2 - 4y_3,$$

$$\frac{dy_2}{dx} = 6y_3 - y_2,$$

$$\frac{dy_3}{dx} = 4y_3 - y_2.$$

3. Prove that the substitution $y_2 = dy/dx, y_3 = d^2y/dx^2, \cdots, y_n = d^{n-1}y/dx^{n-1}$ will reduce the linear nth-order equation

$$\frac{d^n y}{dx^n} + a_1 \frac{d^{n-1}y}{dx^{n-1}} + \cdots + a_{n-1} \frac{dy}{dx} + a_n y = 0$$

to a linear system of the form (11). Write down the matrix of the resulting system and its characteristic equation.

4. Use the method indicated in Problem 3 to solve

$$\frac{d^4 y}{dx^4} - y = 0.$$

42. REAL SYMMETRIC MATRICES

In this section we shall study another important class of matrices that can be reduced to diagonal form by similarity transformations. These are real symmetric matrices. A matrix A is said to be symmetric if $A^t = A$. Since, if A and B are symmetric, $(AB)^t = B^t A^t = BA$, it follows that the product of two symmetric matrices is symmetric if and only if the two matrices commute. In particular, any power of a symmetric matrix is symmetric.

THEOREM 7.8. *The characteristic roots of a real symmetric matrix are real.*

Proof. Let A be any real symmetric matrix, and suppose that it has a complex characteristic root $r + is$. We shall prove that $s = 0$. Since the matrix $A - (r + is)I$ is singular, so also is the matrix

$$\begin{aligned} B &= [A - (r + is)I][A - (r - is)I] \\ &= A^2 - 2rA + (r^2 + s^2)I \\ &= (A - rI)^2 + s^2 I. \end{aligned}$$

Since B is real and singular, there exists a real nonzero vector X such that $BX^t = O$ and therefore $XBX^t = 0$. Hence

$$\begin{aligned} 0 = XBX^t &= X(A - rI)^2 X^t + s^2 XX^t \\ &= X(A - rI)(A - rI)^t X^t + s^2 XX^t \\ &= Y \cdot Y + s^2 X \cdot X, \end{aligned}$$

where Y is the vector $X(A - rI)$. Now, since $X \neq 0$, $X \cdot X > 0$ and since $Y \cdot Y \geq 0$, the above equation implies $s = 0$. The characteristic root $r + is$ is therefore real.

THEOREM 7.9. If A is a real symmetric matrix, there exists an orthogonal matrix P such that $P^{-1}AP$ is a diagonal matrix.

Proof. The proof is a simple modification of that of Theorem 7.7. Let $k_1, k_2, \cdots, k_n$ be the characteristic roots of A. Since k_1 is real, there is a real unit eigenvector S_1 corresponding to k_1. By Theorem 1.17 there exists an orthogonal matrix S of which S_1^t is the first column vector. The first column vector of AS is $AS_1^t = k_1 S_1^t$, and hence the first column vector of $S^{-1}AS$ is $k_1 S^{-1} S_1^t$. This is the first column of $k_1 S^{-1} S$, namely,

$$(k_1, 0, \cdots, 0)^t.$$

Moreover $S^{-1}AS$ is symmetric since $(S^{-1}AS)^t = (S^t AS)^t = S^t A^t S = S^{-1}AS$. Therefore

$$S^{-1}AS = \begin{pmatrix} k_1 & O \\ O & A_1 \end{pmatrix},$$

where A_1 is symmetric of order $n - 1$, and has characteristic roots $k_2, k_3, \cdots, k_n$. The proof is now completed by induction as in the proof of Theorem 7.7. If Q is an orthogonal matrix of order $n - 1$ that diagonalizes A_1, then SR, where

$$R = \begin{pmatrix} 1 & O \\ O & Q \end{pmatrix},$$

is also orthogonal and will diagonalize A.

COROLLARY. A real symmetric matrix of order n has n mutually orthogonal eigenvectors.

Proof. By Theorem 7.5 the column vectors of P are eigenvectors of A. These column vectors are mutually orthogonal since P is an orthogonal matrix.

Conversely it is clear that, if we can find n mutually orthogonal unit eigenvectors of A, then the matrix P of which these are the column vectors is orthogonal and will transform A to diagonal form. The problem of diagonalizing a real symmetric matrix is therefore reduced to that of finding n mutually orthogonal eigenvectors of A. The two following theorems show how this may be done.

THEOREM 7.10. If two eigenvectors S_1 and S_2 of a real symmetric matrix A correspond to different characteristic roots of A, then S_1 and S_2 are orthogonal.

Proof. Let k_1 and k_2 be distinct characteristic roots of A, and let S_1 and S_2 be corresponding eigenvectors. Since $AS_1^t = k_1 S_1^t$, we have $S_2 A S_1^t = k_1 S_2 S_1^t$. Similarly, since $AS_2^t = k_2 S_2^t$ and $A^t = A$, we have $S_2 A = k_2 S_2$, and hence $S_2 A S_1^t = k_2 S_2 S_1^t$. Hence $k_1 S_2 S_1^t = k_2 S_2 S_1^t$ and since $k_1 \neq k_2$, $S_2 S_1^t = 0$ and S_1 and S_2 are orthogonal.

THEOREM 7.11. *If k occurs exactly p times as a characteristic root of a real symmetric matrix A, then A has p but not more than p mutually orthogonal eigenvectors corresponding to k.*

Proof: By Theorem 7.8 there exists a matrix P such that $P^{-1}AP$ is a diagonal matrix in which k occurs exactly p times in the main diagonal. Hence $P^{-1}AP - kI$ has rank $n - p$. Since $P^{-1}AP - kI = P^{-1}(A - kI)P$ and P and P^{-1} are nonsingular, $A - kI$ also has rank $n - p$. Hence by Theorem 2.5, the solution space of the equations

$$(A - kI)X^t = O$$

has dimension $n - (n - p) = p$, and there are therefore p but not more than p mutually orthogonal unit vectors in this space. These are the p mutually orthogonal eigenvectors of A. They are not, of course, uniquely determined.

It follows from Theorem 7.11 that there is a unique (except for a scalar factor ± 1) unit eigenvector of A corresponding to each simple root of the characteristic equation of A, but p mutually orthogonal unit eigenvectors corresponding to each p-fold root. Since, by Theorem 7.10, eigenvectors corresponding to different characteristic roots are orthogonal to each other, this gives us the n mutually orthogonal unit eigenvectors required to construct the matrix P that will transform A to diagonal form.

A real symmetric matrix that has only positive characteristic roots is called a *positive definite* symmetric matrix. The reason for this terminology will appear later, in our study of quadratic forms, but it is useful to introduce the term here.

THEOREM 7.12. *If the matrix of a linear transformation τ of $\mathscr{V}_n(\mathscr{R})$ is positive definite symmetric, then τ is a magnification.*

Proof. Let A be the matrix of τ and P an orthogonal matrix that transforms A to diagonal form. Applying a transformation of coordinates with matrix P, the matrix of τ relative to the new basis will be $P^{-1}AP$, or the diagonal form of A. Since the diagonal elements are the characteristic roots of A and therefore all of them are positive, τ is a magnification.

If A is the matrix of τ relative to a normal orthogonal basis, then the new basis is also normal orthogonal since P is orthogonal. The magnification then takes place in mutually orthogonal directions.

Similarity and Diagonalization Theorems **147**

THEOREM 7.13. A real symmetric matrix A is positive definite if and only if there exists a nonsingular matrix Q such that $A = Q^t Q$.

Proof. If A is real symmetric and there exists a nonsingular matrix Q such that $A = Q^t Q$, then by Theorem 7.9 there exists an orthogonal matrix P such that

$$P^{-1}Q^t QP = \begin{pmatrix} k_1 & & & 0 \\ & k_2 & & \\ & & \ddots & \\ 0 & & & k_n \end{pmatrix}$$

where $k_1, k_2, \cdots, k_n$ are the characteristic roots of $Q^t Q$. But $P^{-1}Q^t QP = P^t Q^t QP = (QP)^t QP$. The ith diagonal element r in this product is therefore equal to the inner product of the ith column vector of QP with itself. This inner product is certainly non-negative. Moreover, it cannot be zero since this would imply that QP has a zero column vector, which is impossible since Q and P are both nonsingular. It follows that $k_1, k_2, \cdots, k_n$ are positive, and, therefore, A is positive definite symmetric.

Conversely, if A is positive definite symmetric, we have an orthogonal matrix P such that

$$P^t AP = \begin{pmatrix} r_1 & & & 0 \\ & r_2 & & \\ & & \ddots & \\ 0 & & & r_n \end{pmatrix} = D,$$

where $r_1, r_2, \cdots, r_n$ are positive. Let D_1 be the diagonal matrix whose diagonal elements are $\sqrt{r_1}, \sqrt{r_2}, \cdots, \sqrt{r_n}$, so that $D_1^2 = D$. Then $A = PDP^t = PD_1 D_1 P^t = (PD_1)(PD_1)^t$, since $D_1^t = D_1$. Since P and D_1 are both nonsingular, $Q^t = PD_1$ is nonsingular, and $A = Q^t Q$. The proof is therefore complete.

THEOREM 7.14. Every real nonsingular matrix A can be written as a product $A = PS$, where S is a positive definite symmetric matrix and P is orthogonal.

Proof. Since A is nonsingular, it follows from Theorem 7.13 that $A^t A$ is positive definite symmetric. Hence there exists an orthogonal matrix Q such that

$$Q^t A^t A Q = \begin{pmatrix} k_1 & & & & 0 \\ & k_2 & & & \\ & & \cdot & & \\ & & & \cdot & \\ 0 & & & & k_n \end{pmatrix} = D,$$

where $k_1, k_2, \cdots, k_n$ are positive real numbers. Let $p_i = \sqrt{k_i}$ ($i = 1, 2, \cdots, n$), and let D_1 be the diagonal matrix with $p_1, p_2, \cdots, p_n$ as diagonal elements so that $D_1^2 = D$.

Now let $S = Q D_1 Q^t$. Since $(Q D_1 Q^t)^t = Q D_1 Q^t$, S is symmetric. Moreover it is positive definite since it is similar to D_1 and hence has positive characteristic roots $p_1, p_2, \cdots, p_n$. Also $S^2 = Q D_1 Q^t Q D_1 Q^t = Q D Q^t = A^t A$. Now, if we let $P = AS^{-1}$, we can show that P is orthogonal. For $P^t P = (AS^{-1})^t AS^{-1} = (S^{-1})^t A^t AS^{-1} = (S^{-1})^t S^2 S^{-1} = (S^{-1})^t S = S^{-1} S = I$, since S, and therefore S^{-1}, is symmetric. Hence $A = PS$, where P is orthogonal and S is positive definite symmetric.

Theorem 7.14 enables us to determine the geometric nature of any nonsingular linear transformation of $\mathscr{V}_n(\mathscr{R})$. Suppose that τ is any such transformation, and let A be the matrix of τ relative to any normal orthogonal basis. Let $A = PS$, where P is orthogonal and S positive definite symmetric. Let Q be an orthogonal matrix that transforms S to its diagonal form D. We then have

$$Q^{-1} A Q = Q^{-1} PS Q = Q^{-1} P Q Q^{-1} S Q = Q^{-1} P Q D = RD,$$

where R is the orthogonal matrix $Q^{-1} P Q$ and D is a diagonal matrix with only positive diagonal elements.

Thus, if we apply a transformation of coordinates with matrix Q, we go over to a new normal orthogonal basis relative to which the matrix of τ is RD. Consequently τ is equivalent to an orthogonal transformation with matrix R (relative to the new basis) followed by a magnification with matrix D (also relative to the new basis). Since it is known that every orthogonal transformation is either a rotation or a rotation followed by a reflection, we have proved the following result:

THEOREM 7.15. *Every nonsingular linear transformation of* $\mathscr{V}_n(\mathscr{R})$ *is*

Similarity and Diagonalization Theorems

equivalent to either a rotation followed by a magnification or to a rotation followed first by a reflection and then by a magnification.

Exercise 7.4

1. Find mutually orthogonal eigenvectors of the following symmetric matrices and hence reduce each of them to diagonal matrix by a similarity transformation:

(a) $\begin{pmatrix} 4 & 3 \\ 3 & -4 \end{pmatrix}$. (b) $\begin{pmatrix} 1 & 2 \\ 2 & -3 \end{pmatrix}$. (c) $\begin{pmatrix} \frac{5}{4} & -\frac{3\sqrt{3}}{4} \\ -\frac{3\sqrt{3}}{4} & -\frac{1}{4} \end{pmatrix}$.

(d) $\begin{pmatrix} 1 & 2 & 0 \\ 2 & 2 & 2 \\ 0 & 2 & 3 \end{pmatrix}$. (e) $\begin{pmatrix} 10 & 0 & 2 \\ 0 & 6 & 0 \\ 2 & 0 & 7 \end{pmatrix}$.

2. Express each of the following matrices as a product of an orthogonal matrix and a positive definite symmetric matrix, and hence express the corresponding linear transformations in terms of rotations, reflections, and magnifications

(a) $\begin{pmatrix} \frac{27}{10} & -\frac{29}{10} \\ \frac{11}{10} & \frac{3}{10} \end{pmatrix}$ (b) $\begin{pmatrix} 1 & -1 & -1 \\ 1 & 1 & -1 \\ 1 & 1 & 1 \end{pmatrix}$.

3. By suitably modifying the proof of Theorem 7.14, show that every real nonsingular matrix A can be written as a product $S_1 P_1$, where S_1 is positive definite symmetric and P_1 is orthogonal.

4. Prove that the matrices P and S of Theorem 7.14 are uniquely determined by A.

5. Prove that A and A^t have the same characteristic roots. If k_1 and k_2 are distinct characteristic roots of A, prove that any eigenvector of A corresponding to k_1 is orthogonal to any eigenvector of A^t corresponding to k_2. Deduce Theorem 7.10 from this.

6. Deduce Theorem 7.9 from Theorem 7.8 and Problem 5, Exercise 7.2.

CHAPTER EIGHT

Reduction of Quadratic Forms

43. REAL QUADRATIC FORMS

A homogeneous quadratic polynomial in n variables $x_1, \cdots, x_n$ is called a *quadratic form*. It was shown in Section 20 that every quadratic form can be written as a matrix product

$$f = XAX^t = \sum_{i=1}^{n} \sum_{j=1}^{n} a_{ij} x_i x_j,$$

where $X = (x_1, x_2, \cdots, x_n)$, and $A = (a_{ij})$ is a symmetric matrix which is uniquely determined by the form f. The matrix A is called the matrix of the form f, its rank is called the rank of f, and the form is said to be singular or nonsingular according as A is singular or nonsingular. If all the coefficients a_{ij} are real, the form is said to be real. In this case it is usual to assume that $x_1, \cdots, x_n$ take only real values. Thus the vector X varies over $\mathscr{V}_n(\mathscr{R})$, and the form f defines a function whose domain is $\mathscr{V}_n(\mathscr{R})$ and whose range is a subset of the real field, called the *range of values* of the form. For example, if $n = 2$, the form $x_1^2 + x_2^2 = XX^t$ has domain $\mathscr{V}_2(\mathscr{R})$ and its range of values is the non-negative real numbers. In this chapter we shall be concerned mainly with real quadratic forms and their applications although we shall occasionally refer to forms whose coefficients may be arbitrary complex numbers.

Our first object is to study the effect on the real quadratic form XAX^t of a transformation of coordinates in $\mathscr{V}_n(\mathscr{R})$. Let P be the matrix of such a transformation, and let $Y = (y_1, \cdots, y_n)$ be the coordinate vector of X relative to the new basis. By Theorem 5.2, $Y = XP$ or $X = YQ$, where $Q = P^{-1}$. We then have

$$f = XAX^t = YQA(YQ)^t = Y(QAQ^t)Y^t.$$

Reduction of Quadratic Forms

Two real quadratic forms XAX^t and YBY^t are said to be *equivalent* if one can be transformed into the other by a transformation of coordinates in $\mathscr{V}_n(\mathscr{R})$. Two forms are said to be *orthogonally equivalent* if one can be transformed into the other by an orthogonal transformation of coordinates in $\mathscr{V}_n(\mathscr{R})$. It may easily be shown that equivalence and orthogonal equivalence of forms are reflexive, symmetric, and transitive relations. From the above result we may now state

THEOREM 8.1. *Two real quadratic forms XAX^t, XBX^t are equivalent if and only if there exists a real nonsingular matrix Q such that $B = QAQ^t$. These two forms are orthogonally equivalent if and only if there exists an orthogonal matrix Q such that $B = QAQ^t$. In each case Q^{-1} is the matrix of the transformation that takes XAX^t into YBY^t.*

A more useful criterion for orthogonal equivalence may now be deduced from Theorem 8.1.

THEOREM 8.2. *Two real quadratic forms XAX^t and XBX^t are orthogonally equivalent if and only if their matrices have the same characteristic roots and these occur with the same multiplicities.*

Proof. If A and B have characteristic roots $k_1, k_2, \cdots, k_n$, there exist orthogonal matrices P and Q such that

$$PAP^t = QBQ^t = \begin{pmatrix} k_1 & & & 0 \\ & k_2 & & \\ & & \cdot & \\ & & & \cdot \\ 0 & & & k_n \end{pmatrix}$$

Hence $B = Q^{-1}PAP^t(Q^t)^{-1} = (Q^{-1}P)A(Q^{-1}P)^t$, and $Q^{-1}P$ is orthogonal since P and Q are. Hence the forms are orthogonally equivalent by Theorem 8.1. Conversely, if the forms are orthogonally equivalent, then $A = QBQ^t = QBQ^{-1}$. Hence A and B are similar and therefore have the same characteristic roots, occurring with the same multiplicities.

COROLLARY. *Every real quadratic form XAX^t is orthogonally equivalent to the form $k_1x_1^2 + k_2x_2^2 + \cdots + k_nx_n^2$, where $k_1, k_2, \cdots, k_n$ are the characteristic roots of A.*

Proof. Choose an orthogonal matrix P such that $D = PAP^t$ is a diagonal matrix. Then $XDX^t = k_1x_1^2 + k_2x_2^2 + \cdots + k_nx_n^2$.

Linear Algebra for Undergraduates

Theorem 8.3. Every real quadratic form XAX^t is equivalent to the form

(1) $$x_1^2 + \cdots + x_p^2 - x_{p+1}^2 - \cdots - x_r^2,$$

where r is the rank of A and p is the number of positive characteristic roots of A.

Proof. If A has rank r, it has exactly r nonzero characteristic roots $k_1, k_2, \cdots, k_r$. Suppose that $k_1, k_2, \cdots, k_p$ are positive and $k_{p+1}, \cdots, k_r$ are negative. Let D be the diagonal matrix of order n whose diagonal elements are

$$k_1, k_2, \cdots, k_r, 0, \cdots, 0,$$

and let Q be the (real) diagonal matrix of order n whose diagonal elements are

$$\frac{1}{\sqrt{k_1}}, \cdots, \frac{1}{\sqrt{k_p}}, \frac{1}{\sqrt{-k_{p+1}}}, \cdots, \frac{1}{\sqrt{-k_r}}, 1, \cdots, 1.$$

Let P be the orthogonal matrix such that $PAP^t = D$. Then

$$N = QPA(QP)^t = QDQ^t = \begin{pmatrix} I_p & & \\ & -I_{r-p} & \\ & & O \end{pmatrix}$$

where the zero matrix O is present only if $r < n$. Since Q and P are nonsingular, so is QP, and hence a transformation of coordinates with matrix $(QP)^{-1}$ will transform XAX^t into the form

$$XNX^t = x_1^2 + \cdots + x_p^2 - x_{p+1}^2 - \cdots - x_r^2.$$

Theorem 8.4. The two forms

$$f = x_1^2 + \cdots + x_p^2 - x_{p+1}^2 - \cdots - x_r^2,$$
$$g = y_1^2 + \cdots + y_q^2 - y_{q+1}^2 - \cdots - y_s^2$$

are equivalent if and only if $r = s$ and $p = q$.

Proof. If $r = s$ and $p = q$, the forms are identical and therefore equivalent. Conversely, if the forms are equivalent, their matrices A and B are connected by the equation $B = QAQ^t$, where Q is nonsingular. Hence A and B have the same rank and $r = s$. It remains to prove that $p = q$. Suppose that $q < p$, and let

(2) $$y_i = \sum_{j=1}^{n} a_{ij} x_j \qquad (i = 1, 2, \cdots, n)$$

be the equations of the transformation in $\mathscr{V}_n(\mathscr{R})$ which takes g into f so that

(3) $$f(x_1, \cdots, x_r) = g(y_1, \cdots, y_s)$$

becomes an identity in $x_1, \cdots, x_n$ when equations 2 are substituted in the right-hand side. Since $q < p$, we can find, by Theorem 1.1, values of $x_1, \cdots, x_n$, not all zero, such that

$$y_1 = a_{11}x_1 + \cdots + a_{1n}x_n = 0,$$
$$\cdots \cdots \cdots \cdots \cdots$$
$$y_q = a_{q1}x_1 + \cdots + a_{qn}x_n = 0,$$
$$x_{p+1} = 0,$$
$$\cdots$$
$$x_n = 0.$$

Substituting these values of $x_1, \cdots, x_n$ in equation 3, we get

$$x_1^2 + x_2^2 + \cdots + x_p^2 = -y_{q+1}^2 - \cdots - y_s^2,$$

which is impossible since $x_1, \cdots, x_p$ are not all zero. Since the supposition that $p < q$ leads to a similar contradiction, it follows that $p = q$ as stated in the theorem.

The form (1) is called the real canonical form of any equivalent form XAX^t. Theorems 8.3 and 8.4 imply the following:

COROLLARY 1. *Two real quadratic forms are equivalent if and only if they have the same real canonical form.*

It also follows that the rank of a real quadratic form and the number p of positive characteristic roots of its matrix are invariant under arbitrary transformations of coordinates in $\mathscr{V}_n(\mathscr{R})$. The number $r - p$ of negative characteristic roots is also invariant, and so also is the number $s = p - (r - p) = 2p - r$. The latter number, the difference between the number of positive and the number of negative characteristic roots is called the *signature* of the quadratic form. We may therefore state

COROLLARY 2. *Two real quadratic forms are equivalent if and only if they have the same rank and the same signature.*

44. QUADRATIC FORMS OVER THE COMPLEX FIELD

If XAX^t is a quadratic form with complex coefficients a_{ij}, we usually allow X to vary over $\mathscr{V}_n(\mathscr{C})$. The form therefore defines a function whose domain is $\mathscr{V}_n(\mathscr{C})$ and whose range of values is a subset of the complex numbers. A transformation of coordinates in $\mathscr{V}_n(\mathscr{C})$ with matrix P transforms XAX^t as before into $X(QAQ^t)X^t$, where $Q = P^{-1}$. Two quadratic forms XAX^t and XBX^t with complex coefficients are said to be *equivalent under transformation of coordinates in* $\mathscr{V}_n(\mathscr{C})$ if one can be transformed into the other by such a transformation. A necessary and sufficient condition for the equivalence of these forms is therefore the existence of a nonsingular matrix Q such that $B = QAQ^t$. By the result stated in Problem 4, Exercise 3.5, a nonsingular matrix Q can always be found such that

$$QAQ^t = \begin{pmatrix} I_r & O \\ O & O \end{pmatrix},$$

where r is the rank of A. Hence we have

THEOREM 8.5. *Every quadratic form XAX^t with complex coefficients is equivalent, under transformation of coordinates in $\mathscr{V}_n(\mathscr{C})$, to the form*

$$x_1^2 + x_2^2 + \cdots + x_r^2,$$

where r is the rank of A.

Since two forms are certainly not equivalent unless they have the same rank, we have also the following:

COROLLARY. *Two quadratic forms with complex coefficients are equivalent under transformation of coordinates in $\mathscr{V}_n(\mathscr{C})$ if and only if they have the same rank.*

THEOREM 8.6. *A quadratic form XAX^t with complex coefficients is the product of two linear factors if and only if its rank is less than or equal to 2.*

Proof. We shall assume that the form is not identically zero since the theorem is trivial in this case. Suppose first that XAX^t factors and that

$$XAX^t = (a_1x_1 + a_2x_2 + \cdots + a_nx_n)(b_1x_1 + b_2x_2 + \cdots + b_nx_n).$$

If the two factors are linearly independent, we may assume, by renaming the variables and coefficients if necessary, that

$$\begin{vmatrix} a_1 & a_2 \\ b_1 & b_2 \end{vmatrix} \neq 0.$$

Reduction of Quadratic Forms 155

The transformation

$$y_1 = a_1 x_1 + a_2 x_2 + \cdots + a_n x_n,$$
$$y_2 = b_1 x_1 + b_2 x_2 + \cdots + b_n x_n,$$
$$y_3 = x_3,$$
$$\cdots$$
$$y_n = x_n$$

is therefore nonsingular and transforms XAX^t into the equivalent form $y_1 y_2$, which has rank 2. By the corollary to Theorem 8.5, the original form also has rank 2.

If the two linear factors are not linearly independent, we can assume $a_1 \neq 0$. The transformation

$$y_1 = a_1 x_1 + a_2 x_2 + \cdots + a_n x_n,$$
$$y_2 = x_2,$$
$$\cdots$$
$$y_n = x_n$$

is then nonsingular and transforms XAX^t into $(b_1/a_1)y_1^2$, which has rank 1. In this case, therefore, XAX^t has rank 1.

Conversely, if XAX^t is not identically zero but has rank less than or equal to 2, it is equivalent under transformation of coordinates in $\mathscr{V}_n(\mathscr{C})$ to either y_1^2 or $y_1^2 + y_2^2$. Since both of these factor (in the field $\mathscr{C}$) into two linear factors, substitution for y_1 and y_2 in terms of $x_1, \cdots, x_n$ will give a factorization of the original form XAX^t.

Exercise 8.1

1. Write down the matrix of each of the following quadratic forms:
   (a) $x_1^2 - 2x_1 x_2 + 2x_2^2$.
   (b) $x_1 x_2 - x_2^2$.
   (c) $x_1 x_2 + x_2 x_3 + x_3 x_1$.
   (d) $x_1^2 - 2x_2^2 + 5x_3^2 - 2x_1 x_2 + 4x_2 x_3 - 16x_3 x_1$.

2. Write the equations of the transformations of coordinates that will reduce the following to real canonical form:
   (a) $x_1^2 - 4x_1 x_2 + x_2^2$.
   (b) $4x_1^2 + x_2^2 - 8x_3^2 + 4x_1 x_2 - 4x_1 x_3 + 8x_2 x_3$.

3. Prove that the range of values of a real quadratic form is not changed by a transformation of coordinates in $\mathscr{V}_n(\mathscr{R})$.

4. A real quadratic form is said to be *positive definite* if and only if its matrix is positive definite. Prove that a real quadratic form XAX^t is positive definite if and only if it satisfies both the following conditions: (a) its range of values is the non-negative real numbers; (b) $XAX^t = 0$ implies $X = O$.

5. A real quadratic form that satisfies condition (a) in Problem 4 is said to be *non-negative definite* (or sometimes *positive semi-definite*). Prove that XAX^t is non-negative definite if and only if all the characteristic roots of A are non-negative and hence that a nonsingular non-negative definite form is positive definite.

6. A real quadratic form is said to be *indefinite* if its range of values includes both positive and negative numbers. Prove that XAX^t is indefinite if and only if A has both positive and negative characteristic roots.

7. Can an indefinite form be equivalent [under transformation of coordinates in $\mathscr{V}_n(\mathscr{R})$] to a positive definite form? To a positive semi-definite form? Why?

45. CLASSIFICATION OF QUADRIC SURFACES

The general equation of the second degree in x, y, and z has the form

(4) $ax^2 + by^2 + cz^2 + 2hxy + 2gxz + 2fyz + 2lx + 2my + 2nz + d = 0$.

If all the coefficients are real the locus in three-dimensional space represented by this equation is called a *quadric surface*. It is assumed that the reader has some acquaintance with these surfaces but a brief description of them has been included in Appendix 2. Our object here is to show how, by suitable rotation and translation of coordinate axes, equation 4 can be reduced to a suitable standard form from which the nature of the surface can be deduced.

We assume that (4) is the equation of the quadric relative to rectangular coordinate axes OX, OY, and OZ. Let E_1, E_2, and E_3 be unit vectors along OX, OY, and OZ, respectively, so that these three vectors constitute the normal orthogonal basis of $\mathscr{V}_3(\mathscr{R})$ defined by

$$E_1 = (1, 0, 0),$$
$$E_2 = (0, 1, 0),$$
$$E_3 = (0, 0, 1).$$

The second-degree terms of equation 4 constitute a quadratic form in x, y, z whose matrix is

$$H = \begin{pmatrix} a & h & g \\ h & b & f \\ g & f & c \end{pmatrix}$$

Reduction of Quadratic Forms

By the corollary to Theorem 7.9, H has three mutually orthogonal unit eigenvectors which may be chosen to be positively oriented. Let these be

(5)
$$F_1 = (\lambda_1, \mu_1, \nu_1) = \lambda_1 E_1 + \mu_1 E_2 + \nu_1 E_3,$$
$$F_2 = (\lambda_2, \mu_2, \nu_2) = \lambda_2 E_1 + \mu_2 E_2 + \nu_2 E_3,$$
$$F_3 = (\lambda_3, \mu_3, \nu_3) = \lambda_3 E_1 + \mu_3 E_2 + \nu_3 E_3,$$

and choose new coordinate axes OX', OY', OZ' along F_1, F_2, F_3, respectively. By Theorem 7.5 the matrix

$$P = \begin{pmatrix} \lambda_1 & \lambda_2 & \lambda_3 \\ \mu_1 & \mu_2 & \mu_3 \\ \nu_1 & \nu_2 & \nu_3 \end{pmatrix}$$

has the property that $P^{-1}HP$ is a diagonal matrix. But, since P is orthogonal, $P^{-1} = P^t$ and equations 5 may be solved for E_1, E_2, E_3 to give

$$E_1 = \lambda_1 F_1 + \lambda_2 F_2 + \lambda_3 F_3,$$
$$E_2 = \mu_1 F_1 + \mu_2 F_2 + \mu_3 F_3,$$
$$E_3 = \nu_1 F_1 + \nu_2 F_2 + \nu_3 F_3.$$

Hence P is the matrix of the transformation of coordinates from the E-basis to the F-basis, i.e., from the x-, y-, z-axes to the x'-, y'-, z'-axes. Moreover, since P is proper orthogonal, this transformation is a rotation of axes. By Theorem 5.2 the old coordinates of a point are given in terms of the new by equations with matrix $(P^t)^{-1}$ but, since P is orthogonal, $(P^t)^{-1} = P$, and these equations become

(6)
$$x = \lambda_1 x' + \lambda_2 y' + \lambda_3 z',$$
$$y = \mu_1 x' + \mu_2 y' + \mu_3 z',$$
$$z = \nu_1 x' + \nu_2 y' + \nu_3 z'.$$

Now, if we write $X = (x, y, z)$ and $X' = (x', y', z')$, we have $X = X'P^t$, and the second-degree terms of (4) which constitute the quadratic form XHX^t become, on applying our rotation of axes,

$$X'(P^tHP)X'^t = X'(P^{-1}HP)X'^t = X'KX'^t,$$

where

$$K = \begin{pmatrix} k_1 & 0 & 0 \\ 0 & k_2 & 0 \\ 0 & 0 & k_3 \end{pmatrix}$$

and k_1, k_2, k_3 are the characteristic roots of H corresponding to F_1, F_2, F_3, respectively. The quadratic form XHX^t is therefore reduced by the rotation of axes to $k_1 x'^2 + k_2 y'^2 + k_3 z'^2$, and the equation of the quadric now has the form

(7) $\quad k_1 x'^2 + k_2 y'^2 + k_3 z'^2 + 2l' x' + 2m' y' + 2n' z' + d = 0.$

We now investigate the different surfaces that can be represented by equation 7.

Case I (k_1, k_2, k_3 all different from zero). In this case the translation of axes defined by

$$x'' = x' + l'/k_1,$$
$$y'' = y' + m'/k_2,$$
$$z'' = z' + n'/k_3$$

reduces equation 7 to the form

(7') $\quad k_1 x^2 + k_2 y^2 + k_3 z^2 + d' = 0,$

where, for convenience, the double primes have been omitted.

I(a) (k_1, k_2, k_3 all have the same sign):

1. If d' agrees in sign with k_1, the surface contains no real points but will be called an imaginary ellipsoid.
2. If d' differs in sign from k_1, the surface is a real ellipsoid.
3. If $d' = 0$, the surface is an imaginary quadric cone containing one real point $(0, 0, 0)$.

I(b) (k_1, k_2, k_3 not all of the same sign):

4. If sign of d' agrees with the sign of two of the k_i's, the surface is a hyperboloid of two sheets.
5. If the sign of d' differs from that of two of the k_i's, the surface is a hyperboloid of one sheet.
6. If $d' = 0$, the surface is a real quadric cone.

Case II (One of k_1, k_2, k_3 is zero, two different from zero). In this case, assuming for example that $k_3 = 0$, the translation

$$x'' = x' + l'/k_1,$$
$$y'' = y' + m'/k_2,$$
$$z'' = z'$$

reduces equation 7 to the form

$$k_1 x^2 + k_2 y^2 + 2n' z + d' = 0.$$

II(a) If $n' \neq 0$, the translation $z' = z + d'/2n'$ gives $k_1 x^2 + k_2 y^2 + 2n' z = 0$.

7. If k_1, k_2 have like signs, the surface is an elliptic paraboloid.

Reduction of Quadratic Forms 159

8. If k_1, k_2 have opposite signs, the surface is a hyperbolic paraboloid.

II(b) If $n' = 0$, the equation has the form $k_1 x^2 + k_2 y^2 + d' = 0$.

9. If k_1, k_2 have the same sign and d' the opposite sign, the surface is a real elliptic cyclinder.
10. If k_1, k_2 and d' all have the same sign, the surface is an imaginary elliptic cylinder.
11. If k_1, k_2 have opposite signs and $d' \neq 0$, the surface is a hyperbolic cylinder.
12. If k_1, k_2 have the same sign and $d' = 0$, the surface is two imaginary planes intersecting in a real line.
13. If k_1 and k_2 have opposite signs and $d' = 0$, the surface is two real intersecting planes.

Case III (If two of k_1, k_2, k_3 are zero). In this case, assuming $k_1 \neq 0$, $k_2 = k_3 = 0$, equation 7 takes the form $k_1 x^2 + 2m'y + 2n'z + d' = 0$ after the translation $x'' = x' + l'/k_1$, $y'' = y'$, $z'' = z'$.

III(a) (m' and n' not both zero). If m', n' are both different from zero, a suitable rotation about the x-axis will reduce the equation to

$$k_1 x^2 + m''y + d' = 0 \qquad m'' \neq 0,$$

and a translation will remove the constant term.

14. Hence, if m', n' are not both zero, the surface is a parabolic cylinder.

III(b) ($m' = n' = 0$).

15. If k_1 and d' have like signs, then the surface is two imaginary parallel planes.
16. If k_1 and d' differ in sign, the surface is two real parallel planes.
17. If $d' = 0$, the surface is two real coincident planes.

This completes the first classification of the seventeen different types of quadric surfaces. Of the seventeen, numbers 1, 2, 4, 5, 7, and 8 are the six *true quadrics*. Five of these are real and one imaginary. All the rest are called *singular* quadrics. The reason for this distinction will appear in the next section where we shall give another classification of quadrics according to rank.

Exercise 8.2

1. Reduce the following equations to standard form by a suitable rotation of axes followed if necessary by a translation. Hence determine the type of quadric that each equation represents.

   (a) $2xy + 2yz + 2xz = 1$.
   (b) $2x^2 + z^2 + 2xy + 3xz + yz - 2 = 0$.

(c) $2x^2 + 2y^2 + 5z^2 - 4xy + 2yz - 2xz - 10x - 6y - 2z = 0$.

(d) $x^2 - yz + 2x - 4y + 3z = 12$.

2. Show that equation 4 represents a surface of revolution if and only if the matrix H has two equal nonzero characteristic roots.

3. Use the theory of the orthogonal reduction of quadratic forms to find the equations for a rotation of axes that will reduce the equation of the conic

$$x^2 - 4xy + 4y^2 + 2x - 2y + 6 = 0$$

to standard form. Determine what type of curve this is, and sketch its graph.

4. Classify all curves, real or imaginary, which can be represented by the general equation

$$ax^2 + 2hxy + by^2 + 2gx + 2fy + c = 0,$$

according to the nature of the characteristic roots of the matrix

$$\begin{pmatrix} a & h \\ h & b \end{pmatrix}$$

Hence obtain the usual criteria for the equation to represent an ellipse, hyperbola, parabola, or circle.

5. Find a necessary and sufficient condition on the coefficients of the equation in Problem 4 that this equation represent two (real or imaginary) straight lines. *Hint:* show that the left-hand side of this equation factors if and only if the quadratic form $ax^2 + 2hxy + by^2 + 2gxt + 2fyt + ct^2$ factors, and then use Theorem 8.6.

46. CLASSIFICATION OF QUADRICS BY RANK

Consider the general quadric defined by equation 4 and denote the left-hand side of this equation by $\varphi(x, y, z)$. We define the associated quadratic form

$$\psi(x, y, z, u) = u^2 \varphi(x/u, y/u, z/u)$$

$= ax^2 + by^2 + cz^2 + 2hxy + 2gxz + 2fyz + 2lxu + 2myu + 2nzu + du^2$

whose matrix is

$$D = \begin{pmatrix} a & h & g & l \\ h & b & f & m \\ g & f & c & n \\ l & m & n & d \end{pmatrix}$$

The determinant $|D|$ is called the *discriminant* of the quadric, and the rank of D is called the rank of the quadric. It is easy to see that $\varphi(x, y, z)$ factors into (real or imaginary) linear factors if and only if $\psi(x, y, z, u)$

Reduction of Quadratic Forms

does. By Theorem 8.6 this is the case if and only if D has rank ≤ 2. It follows that a quadric consists of two planes (real or imaginary, distinct or coincident) if and only if its rank ≤ 2. It is not difficult to show that, if the rank is 2, the two planes are distinct but that, if the rank is 1, they are coincident.

In order to distinguish the quadrics of rank 3 from those of rank 4, we apply to $\psi(x, y, z, u)$ the transformation

(8)
$$x = \lambda_1 x' + \lambda_2 y' + \lambda_3 z',$$
$$y = \mu_1 x' + \mu_2 y' + \mu_3 z',$$
$$z = \nu_1 x' + \nu_2 y' + \nu_3 z',$$
$$u = u'$$

whose matrix is $Q = \begin{pmatrix} P & 0 \\ 0 & 1 \end{pmatrix}$. Here P is the matrix of Section 45 whose columns are unit eigenvectors of H. Writing X and X' for the vectors (x, y, z, u) and (x', y', z', u'), the form $\psi(x, y, z, u) = XDX^t$ is transformed by (8) into $X'D'X'^t$, where

$$D' = Q^t D Q = \begin{pmatrix} k_1 & 0 & 0 & l' \\ 0 & k_2 & 0 & m' \\ 0 & 0 & k_3 & n' \\ l' & m' & n' & d \end{pmatrix}$$

and k_1, k_2, k_3 are as before the characteristic roots of H. Since Q is nonsingular, the rank of D' is equal to that of D. If k_1, k_2, k_3 are all different from 0, D' is equivalent under elementary transformations to the diagonal matrix

$$D'' = \begin{pmatrix} k_1 & 0 & 0 & 0 \\ 0 & k_2 & 0 & 0 \\ 0 & 0 & k_3 & 0 \\ 0 & 0 & 0 & d' \end{pmatrix},$$

in which $d' = d - l'^2/k_1 - m'^2/k_2 - n'^2/k_3$. Hence D'' is the matrix associated with equation 7'. Since its rank is the same as that of D, we conclude that the two ellipsoids and the two hyperboloids have rank 4 but that the quadric cones ($d' = 0$) have rank 3.

Turning to Case II, if k_1 and k_2 are different from 0 but $k_3 = 0$, D' is equivalent to

$$\begin{pmatrix} k_1 & 0 & 0 & 0 \\ 0 & k_2 & 0 & 0 \\ 0 & 0 & 0 & n' \\ 0 & 0 & n' & d' \end{pmatrix},$$

which has rank 4 if $n' \neq 0$ but rank 3 if $n' = 0$ and $d' \neq 0$. The two paraboloids therefore have rank 4 and the elliptic and hyperbolic cylinders rank 3.

Continuing thus we can easily complete the following classification of the seventeen quadrics by rank:

1. Quadrics of rank 4: real and imaginary ellipsoids, the two hyperboloids, and the two paraboloids.
2. Quadrics of rank 3: real and imaginary cones; the elliptic, hyperbolic, and parabolic cylinders.
3. Quadrics of rank 2: two intersecting planes (real or imaginary); two parallel planes (real or imaginary).
4. Quadric of rank 1: two coincident planes.

A quadric of rank 4 is called a *true* or *nonsingular* quadric. A quadric of rank less than 4 is called a *singular* quadric, and one of rank less than 3, a *degenerate* quadric.

47. INVARIANTS

In this section we shall show that certain functions of the coefficients of equation 4 are invariant under rotation and translation of axes.

THEOREM 8.7. *Under a rotation of axes the matrices H and D associated with equation 4 are replaced by similar matrices.*

Proof. If P is the matrix of the rotation, we have already seen that H is replaced by $P^tHP = P^{-1}HP$, which is similar to H.

Suppose now that the rotation, with matrix P, defined by equations 6 transforms (4) into

(9) $$\Phi(x', y', z') = 0,$$

so that

$$\Phi(x', y', z') = \varphi(x, y, z).$$

Reduction of Quadratic Forms

The quadratic forms associated with (4) and (9) are then

(10) $$u^2\varphi(x/u, y/u, z/u)$$

and

(11) $$u'^2\Phi(x'/u', y'/u', z'/u'),$$

and we see that (10) is transformed into (11) by the transformation defined by equations 8. The matrix Q of this transformation is orthogonal since P is. Hence the matrix D is replaced under the rotation by $Q^t DQ = Q^{-1}DQ$, which is similar to D.

THEOREM 8.8. The following functions of the coefficients of equation 4 are invariant under rotation of axes:

1. $$|H| = \begin{vmatrix} a & h & g \\ h & b & f \\ g & f & c \end{vmatrix}$$

2. $j = ab + bc + ca - h^2 - f^2 - g^2$.

3. $k = a + b + c$.

4. $$|D| = \begin{vmatrix} a & h & g & l \\ h & b & f & m \\ g & f & c & n \\ l & m & n & d \end{vmatrix}.$$

5. $q = \begin{vmatrix} a & h & g \\ h & b & f \\ g & f & c \end{vmatrix} + \begin{vmatrix} a & h & l \\ h & b & m \\ l & m & d \end{vmatrix} + \begin{vmatrix} a & g & l \\ g & c & n \\ l & n & d \end{vmatrix} + \begin{vmatrix} b & f & m \\ f & c & n \\ m & n & d \end{vmatrix}.$

6. $r = ab + bc + ca + ad + bd + cd - f^2 - g^2 - h^2 - l^2 - m^2 - n^2$.

7. $s = a + b + c + d$.

Proof. Since a rotation of axes replaces D and H by similar matrices, the characteristic polynomials $|D - xI|$ and $|H - xI|$ are invariant under rotation of axes. The proof of the theorem, therefore, consists in noting that $|H|$, $-j$, and k are, respectively, the constant term, the coefficient of x, and the coefficient of x^2 in $|H - xI|$ and that $|D|$, $-q$, r, and $-s$ are, respectively, the constant term and the coefficients of x, x^2, and x^3 in $|D - xI|$.

THEOREM 8.9. The expressions $|D|$, $|H|$, j, k are invariant under translation of axes.

Proof. Since the translation

(12)
$$x = x' + \alpha,$$
$$y = y' + \beta,$$
$$z = z' + \gamma,$$

when applied to equation 4, leaves the coefficients of all second-degree terms in x, y, z unchanged, it follows that the matrix H is unchanged. Hence $|H|$, j and k are invariant under translation. To prove that $|D|$ is also invariant, we must investigate the effect of this translation on the quadratic form (10). Under the translation defined by (12) equation 4 becomes

$$\varphi(x' + \alpha, y' + \beta, z' + \gamma) = 0,$$

and the associated quadratic form is

$$u'^2 \varphi\left(\frac{x' + \alpha u'}{u'}, \frac{y' + \beta u'}{u'}, \frac{z' + \gamma u'}{u'}\right).$$

Comparing this with $u^2\varphi(x/u, y/u, z/u)$, we see that under the translation (12) the associated quadratic form undergoes the transformation

$$x = x' + \alpha u',$$
$$y = y' + \beta u',$$
$$z = z' + \gamma u',$$
$$u = u',$$

whose matrix is

$$C = \begin{pmatrix} 1 & 0 & 0 & \alpha \\ 0 & 1 & 0 & \beta \\ 0 & 0 & 1 & \gamma \\ 0 & 0 & 0 & 1 \end{pmatrix}.$$

Under this translation, therefore, D is replaced by $C^t DC = D'$, and $|D'| = |C^t| \cdot |D| \cdot |C| = |D| \cdot |C|^2 = |D|$, since $|C| = 1$.

Exercise 8.3

1. Prove that the following functions of the coefficients of equation 4 are invariant under rotation of axes:
  (a) $u = l^2 + m^2 + n^2$.
  (b) $v = 2fmn + 2gln + 2hlm + al^2 + bm^2 + cn^2$.

Reduction of Quadratic Forms

2. Construct examples to show that q, r, and s (defined in Theorem 8.8) are not invariant under arbitrary translation of axes.

3. Prove that a nonsingular quadric is a paraboloid if and only if $|H| = 0$.

4. Prove that equation 4 represents an elliptic paraboloid if and only if $|D| < 0$, $|H| = 0$, and $j > 0$.

5. Derive necessary and sufficient conditions that equation 4 represent a hyperbolic paraboloid.

6. Prove that equation 4 represents a real ellipsoid if and only if its coefficients satisfy either

(a) $|D| < 0$, $|H| > 0$, $j > 0$, and $k > 0$

or

(b) $|D| < 0$, $|H| < 0$, $j > 0$, and $k < 0$.

7. Prove that for a hyperboloid of one sheet $|D| > 0$ but for a hyperboloid of two sheets $|D| < 0$.

8. Prove that, if $|D| > 0$, $|H| < 0$, and $k > 0$, equation 4 represents a hyperboloid of one sheet.

9. Derive necessary and sufficient conditions that equation 4 represent a hyperboloid of one sheet.

48. A PROBLEM IN DYNAMICS

We shall now prove an important theorem on the simultaneous reduction of two quadratic forms and apply it to a problem in dynamics.

THEOREM 8.10. Suppose that $X \in \mathscr{V}_n(\mathscr{R})$, and let $f = XAX^t$ and $g = XBX^t$ be two real quadratic forms one of which, say f, is positive definite. There exists a transformation of coordinates in $\mathscr{V}_n(\mathscr{R})$ that reduces f and g, respectively, to the forms

$$f' = x_1'^2 + x_2'^2 + \cdots + x_n'^2$$

and

$$g' = k_1 x_1'^2 + k_2 x_2'^2 + \cdots + k_n x_n'^2,$$

where $k_1, k_2, \cdots, k_n$ are the roots of the determinantal equation

$$|B - xA| = |b_{ij} - a_{ij}x| = 0.$$

The coefficients $k_1, k_2, \cdots, k_n$ are positive if and only if XBX^t is positive definite.

Proof. Since XAX^t is positive definite, it is equivalent, by Theorem 8.3, to the form

$$f' = x_1'^2 + x_2'^2 + \cdots + x_n'^2.$$

Let P be the matrix of the transformation that effects this reduction, so that $P^{-1}A(P^{-1})^t = I$ or $A = PP^t$. The same transformation reduces g to $g'' = X'P^{-1}B(P^{-1})^tX'^t$. Since $P^{-1}B(P^{-1})^t$ is a real symmetric matrix, there exists an orthogonal transformation with matrix Q that reduces g'' to the form

$$g' = k_1 x_1'^2 + k_2 x_2'^2 + \cdots + k_n x_n'^2,$$

where $k_1, \cdots, k_n$ are the characteristic roots (necessarily real) of $P^{-1}B(P^{-1})^t$. Being orthogonal, this transformation leaves f' unchanged. Hence the product transformation with matrix PQ will effect the required simultaneous reduction. Moreover, since $k_1, k_2, \cdots, k_n$ are the characteristic roots of $P^{-1}B(P^{-1})^t$, they are also those of the similar matrix

$$P[P^{-1}B(P^{-1})^t]P^{-1} = B(P^{-1})^tP^{-1} = B(PP^t)^{-1} = BA^{-1}.$$

Hence they are the roots of the equation

$$|BA^{-1} - xI| = 0,$$

or, since $|A| \neq 0$, of the equivalent equation

$$|B - xA| = 0.$$

Finally, by Theorem 8.4, Corollary 2, the rank and signature of g' are equal to the rank and signature of g. Hence a necessary and sufficient condition that $k_1, k_2, \cdots, k_n$ are all positive is that the form XBX^t be positive definite.

We shall apply Theorem 8.10 to the solution of the problem of small oscillations of a conservative dynamical system about a position of stable equilibrium. The student who is not familiar with Lagrange's equations in generalized coordinates is referred to [12] or [16] for the derivation of these. In [16] he will also find a more extensive discussion of the theory of vibrations.

We assume that the position of the vibrating system is specified at any time by n generalized coordinates $q_1, q_2, \cdots, q_n$ whose derivatives with respect to time are denoted by $\dot{q}_1, \dot{q}_2, \cdots, \dot{q}_n$. We assume also that the equilibrium position is defined by $q_1 = q_2 = \cdots = q_n = 0$. The kinetic energy of the system is then a quadratic form

$$T = \sum_{i=1}^{n} \sum_{j=1}^{n} a_{ij} \dot{q}_i \dot{q}_j$$

in the velocities $\dot{q}_i$. Although the coefficients a_{ij} in general depend on the coordinates, if the oscillations are small an approximation to the motion can be obtained by assuming a_{ij} to be constant. We shall make this assumption.

Reduction of Quadratic Forms

Lagrange's equations of motion for a conservative system are

$$\text{(13)} \qquad \frac{d}{dt}\left(\frac{\partial T}{\partial \dot{q}_i}\right) - \frac{\partial T}{\partial q_i} = -\frac{\partial V}{\partial q_i},$$

where V is the potential energy of the system. Let V be expanded in a Taylor's series about the point $q_i = 0$ ($i = 1, 2, \cdots, n$). We may assume that the constant term is 0 since it will not affect equations 13. The first-degree terms in q_1 also are absent since V has a minimum value at a point of stable equilibrium, and therefore $\partial V/\partial q_i = 0$ ($i = 1, 2, \cdots, n$), at the point $q_1 = q_2 = \cdots = q_n = 0$. Hence by neglecting cubic and higher terms, since the q_i are small, we can approximate V by a quadratic form

$$V = \sum_{i=1}^{n} \sum_{j=1}^{n} b_{ij} q_i q_j$$

in the coordinates q_i. The coefficients b_{ij} are constants.

Under the above assumptions concerning the kinetic energy, equations 13 take the form

$$\text{(14)} \qquad \frac{d}{dt}\left(\frac{\partial T}{\partial \dot{q}_i}\right) = -\frac{\partial V}{\partial q_i}.$$

If the coordinates q_i are subjected to a transformation,

$$\text{(15)} \qquad q_i = \sum_{j=1}^{n} p_{ij} q'_j,$$

where p_{ij} are constants, it follows, on differentiation with respect to time, that

$$\dot{q}_i = \sum_{j=1}^{n} p_{ij} \dot{q}'_j,$$

and hence the velocities $\dot{q}_i$ are subjected to the same transformation. By the nature of kinetic energy the quadratic form T is positive definite (see Problem 4, Exercise 8.1). Hence, by Theorem 8.10, the transformation (15) can be chosen so that T and V are transformed, respectively, into

$$T = \dot{q}'^2_1 + \dot{q}'^2_2 + \cdots + \dot{q}'^2_n$$

and

$$V = k_1 q'^2_1 + k_2 q'^2_2 + \cdots + k_n q'^2_n.$$

Moreover, since V has a minimum value 0 at the point $q'_1 = q'_2 = \cdots = q'_n = 0$, it follows that V is positive definite and $k_i > 0$ ($i = 1, 2, \cdots, n$). We can therefore write $k_i = \omega_i^2$, where $\omega_1, \cdots, \omega_n$ are real and positive.

Substituting in (14), the equations of motion now become

$$\frac{d^2 q_i'}{dt^2} = -\omega_i^2 q_i' \qquad (i = 1, 2, \cdots, n),$$

and their solutions are

(16) $\qquad q_i' = a_i \cos(\omega_i t + b_i) \qquad (i = 1, 2, \cdots, n).$

The coordinates $q_1', \cdots, q_n'$ are called *normal coordinates* for the vibrating system. Equations 16 imply that each normal coordinate is subject to a simple harmonic vibration.

• The orthogonal reduction of quadratic forms has applications such as the one given above in many fields of applied mathematics. It is a procedure that is often necessary in mathematical statistics in discussing distribution functions of several correlated variables. A brief summary of such statistical applications will be found in [15].

CHAPTER NINE

Vector Spaces over the Complex Field

49. INNER PRODUCTS IN $\mathscr{V}_n(\mathscr{C})$

The student has probably noticed that a considerable portion of the theory of real vector spaces covered in the preceding chapters has stemmed from our definition of the inner product of two vectors. First, the introduction of the inner product enabled us to define length and orthogonality of vectors. This led to the discussion of normal orthogonal bases in $\mathscr{V}_n(\mathscr{R})$ and its subspaces and to the concept of orthogonal transformations, i.e., linear transformations of $\mathscr{V}_n(\mathscr{R})$ that leave inner products invariant. In this chapter we shall develop an analogous theory for $\mathscr{V}_n(\mathscr{C})$. Many of the theorems and their proofs are so similar to those previously obtained for real vector spaces that we shall omit their proofs.

If x is a complex number, we write $\bar{x}$ for the complex conjugate of x and $|x|$ for the absolute value of x. Thus $|x|^2 = x\bar{x}$. Similarly, if $A = (a_{ij})$ is a matrix with complex elements, $\bar{A}$ is the matrix $(\bar{a}_{ij})$ in which each element of A is replaced by its complex conjugate. The student should verify that $\overline{AB} = \bar{A}\bar{B}$, $\overline{A+B} = \bar{A} + \bar{B}$, and $(\bar{A})^{-1} = \overline{A^{-1}}$. The transposed conjugate of A, namely, $\bar{A}^t$, is given a special name owing to its frequent occurrence. It is called the *adjoint* of A.

DEFINITION. If X and Y are vectors of $\mathscr{V}_n(\mathscr{C})$, their inner product $X \cdot Y$ is defined to be

$$X \cdot Y = X\bar{Y}^t = x_1\bar{y}_1 + x_2\bar{y}_2 + \cdots + x_n\bar{y}_n.$$

We note that, if X and Y are real, $\bar{Y} = Y$ and this definition reduces to the previous one. If X, Y, and Z are any vectors of $\mathscr{V}_n(\mathscr{C})$ and c is any complex number, the following laws of inner multiplication can easily be proved from the definition:

IMC1 $X \cdot X \geq 0$ and $X \cdot X = 0$ if and only if $X = O$.
IMC2 $X \cdot Y = \overline{Y \cdot X}$.
IMC3 $(cX) \cdot Y = c(X \cdot Y)$,
       $X \cdot (cY) = \bar{c}(X \cdot Y)$.
IMC4 $X \cdot (Y + Z) = X \cdot Y + X \cdot Z$,
       $(Y + Z) \cdot X = Y \cdot X + Z \cdot X$.

Guided by analogy with the real case, we now define the *length* of a vector X of $\mathscr{V}_n(\mathscr{C})$ to be $\|X\| = \sqrt{X \cdot X}$, which by IMC1 is a non-negative real number. Also two vectors X and Y are *orthogonal* to each other if $X \cdot Y = 0$. (Note that, by IMC2, $Y \cdot X = 0$ if and only if $X \cdot Y = 0$.)

THEOREM 9.1. If X and Y are arbitrary vectors of $\mathscr{V}_n(\mathscr{C})$, then

(a) $\|X\| \geq 0$ and $\|X\| = 0$ if and only if $X = O$.

(b) $\|cX\| = |c|\ \|X\|$ for any complex number c.

(c) $|X \cdot Y| \leq \|X\|\ \|Y\|$ (the Schwarz inequality).

(d) $\|X + Y\| \leq \|X\| + \|Y\|$ (the triangle inequality).

Proof. Parts (a) and (b) follow directly from the definition of inner product.

We note that (c) holds with the sign of equality if either $X = O$ or $Y = O$. We assume therefore that X and Y are nonzero vectors. By IMC1 through 4 we have, for all real values of λ,

$$(\lambda X + Y) \cdot (\lambda X + Y) = \lambda^2 X \cdot X + \lambda(X \cdot Y + Y \cdot X) + Y \cdot Y$$
$$= \|X\|^2 \lambda^2 + 2R(X \cdot Y)\lambda + \|Y\|^2 \geq 0,$$

where $R(X \cdot Y)$ denotes the real part of the complex number $X \cdot Y$. Since a quadratic function of λ is non-negative only if its discriminant is non-positive, we conclude that

$$[R(X \cdot Y)]^2 - \|X\|^2 \|Y\|^2 \leq 0$$

or

(1) $R(X \cdot Y) \leq \|X\|\ \|Y\|$.

Since (1) holds for all X and Y, we can replace Y by $e^{i\theta}Y$, whence, by IMC3 and part (b),

$$R[e^{-i\theta}(X \cdot Y)] \leq \|X\|\ \|Y\|.$$

Now choose θ to be the amplitude of the complex number $X \cdot Y$ so that $e^{-i\theta}(X \cdot Y)$ is real and $R[e^{-i\theta}(X \cdot Y)] = |X \cdot Y|$. Hence for all X and Y

$$|X \cdot Y| \leq \|X\|\ \|Y\|.$$

Vector Spaces over the Complex Field

The triangle inequality now follows. Using IMC1 through 4 and the Schwarz inequality, we have

$$(X + Y)\cdot(X + Y) = X\cdot X + X\cdot Y + Y\cdot X + Y\cdot Y$$
$$= \|X\|^2 + 2R(X\cdot Y) + \|Y\|^2$$
$$\leq \|X\|^2 + 2|X\cdot Y| + \|Y\|^2$$
$$\leq \|X\|^2 + 2\|X\|\,\|Y\| + \|Y\|^2$$
$$= (\|X\| + \|Y\|)^2.$$

Hence, taking positive square roots,

$$\|X + Y\| \leq \|X\| + \|Y\|$$

as required.

The above definition of orthogonality enables us to extend the results of Theorems 1.13 through 1.17 to $\mathscr{V}_n(\mathscr{C})$. In fact the proofs of these theorems given in Section 10 apply to the new situation with only occasional modifications necessitated by the difference between the rules IMI1 through 4 of Section 10 and the corresponding rules IMCI through 4 for inner products in $\mathscr{V}_n(\mathscr{C})$. For convenience we restate the main results of these theorems for $\mathscr{V}_n(\mathscr{C})$, but the proofs will not be repeated.

THEOREM 9.2. (a) Any set of nonzero mutually orthogonal vectors of $\mathscr{V}_n(\mathscr{C})$ are linearly independent.
  (b) If $\mathscr{S}$ is a proper subspace of $\mathscr{V}_n(\mathscr{C})$, there exists a nonzero vector of $\mathscr{V}_n(\mathscr{C})$ that is orthogonal to $\mathscr{S}$, i.e., to every vector in $\mathscr{S}$.
  (c) Every subspace of $\mathscr{V}_n(\mathscr{C})$ has a normal orthogonal basis, i.e., a basis consisting of mutually orthogonal unit vectors.
  (d) Let $\mathscr{S}$ be a subspace of $\mathscr{V}_n(\mathscr{C})$ of dimension r, and suppose that $E_1, \cdots, E_s$ ($1 \leq s < r$) are s mutually orthogonal unit vectors in $\mathscr{S}$. There exist unit vectors $E_{s+1}, \cdots, E_r$ in $\mathscr{S}$ such that $E_1, \cdots, E_r$ is a normal orthogonal basis of $\mathscr{S}$.

If $\mathscr{S}$ is a subspace of $\mathscr{V}_n(\mathscr{C})$, the set $\mathscr{T}$ of all vectors of $\mathscr{V}_n(\mathscr{C})$ that are orthogonal to $\mathscr{S}$ is called the orthogonal complement of $\mathscr{S}$ in $\mathscr{V}_n(\mathscr{C})$. The results of Theorems 4.1 through 4.3 can now be carried over to $\mathscr{V}_n(\mathscr{C})$ with the obvious minor modifications. We state the main results without further proof.

THEOREM 9.3. If $\mathscr{T}$ is the orthogonal complement of $\mathscr{S}$ in $\mathscr{V}_n(\mathscr{C})$,
  (a) $\mathscr{S} \cap \mathscr{T} = 0$, $\mathscr{S} + \mathscr{T} = \mathscr{V}_n(\mathscr{C})$, and $\mathscr{T}$ has dimension $n - r$, where r is the dimension of $\mathscr{S}$.

(b) Every vector X of $\mathscr{V}_n(\mathscr{C})$ has a unique representation $X = S + T$, where $S \in \mathscr{S}$ and $T \in \mathscr{T}$. The vectors S and T are called the *orthogonal projections*, of X on the spaces $\mathscr{S}$ and $\mathscr{T}$, respectively.

(c) $\mathscr{S}$ is the orthogonal complement of $\mathscr{T}$ in $\mathscr{V}_n(\mathscr{C})$.

The concept of volume of a parallelepiped can also be extended to $\mathscr{V}_n(\mathscr{C})$. Let $X_1, \cdots, X_r, r \leq n$, be any r vectors of $\mathscr{V}_n(\mathscr{C})$. We define the r-volume of the r-dimensional parallelepiped Π defined by $X_1, \cdots, X_r$ by induction on r. For $r = 1$, the 1-volume of X_1 is its length, namely, $\sqrt{X \cdot X}$. Assuming that $(r - 1)$-volume has been defined, the r-volume of Π is equal to the product of the $(r - 1)$-volume of the parallelepiped defined by $X_1, \cdots, X_{r-1}$ and the length of the orthogonal projection of X_r onto the orthogonal complement in $\mathscr{V}_n(\mathscr{C})$ of the space generated by $X_1, \cdots, X_{r-1}$. Since the length of a vector is real and non-negative, the r-volume of any parallelepiped in $\mathscr{V}_n(\mathscr{C})$ is also real and non-negative. It now follows, as in Theorem 4.7, that the r-volume of the parallelepiped defined by $X_1, \cdots, X_r$ is the positive square root of the determinant

(2)
$$\begin{vmatrix} X_1 \cdot X_1 & X_1 \cdot X_2 & \cdots & X_1 \cdot X_r \\ X_2 \cdot X_1 & X_2 \cdot X_2 & \cdots & X_2 \cdot X_r \\ \cdot & \cdot & \cdot & \cdot \\ X_r \cdot X_1 & X_r \cdot X_2 & \cdots & X_r \cdot X_r \end{vmatrix}.$$

This determinant is therefore real and non-negative and is zero if and only if $X_1, \cdots, X_r$ are linearly dependent. (See Problem 8, Exercise 4.3.)

Exercise 9.1

1. Prove that in the Schwartz inequality $|X \cdot Y| \leq \|X\| \|Y\|$ the sign of equality holds if and only if X and Y are linearly dependent. Is this true of the triangle inequality?

2. Prove that in $\mathscr{V}_n(\mathscr{C})$ every subspace of dimension r is the solution space of a set of $n - r$ independent equations.

3. Give an algebraic proof, independent of the definition of volume, that the determinant (2) is a real number.

4. Prove that the vanishing of the determinant (2) is a necessary and sufficient condition for the linear dependence of $X_1, X_2, \cdots, X_r$.

50. NORMAL ORTHOGONAL BASES AND UNITARY TRANSFORMATIONS

Let $E_1, \cdots, E_n$ and $F_1, \cdots, F_n$ be two normal orthogonal bases of $\mathscr{V}_n(\mathscr{C})$, and let $U = (u_{ij})$ be the matrix of the transformation of coordinates

Vector Spaces over the Complex Field

from the E-basis to the F-basis. Since $E_i \cdot E_j = F_i \cdot F_j = \delta_{ij}$, and

$$E_i = \sum_{j=1}^{n} u_{ij} F_j \qquad (i = 1, 2, \cdots, n),$$

we have, from IMC2 through 4,

$$E_i \cdot E_j = (u_{i1}F_1 + u_{i2}F_2 + \cdots + u_{in}F_n) \cdot (u_{j1}F_1 + u_{j2}F_2 + \cdots + u_{jn}F_n)$$
$$= u_{i1}\bar{u}_{j1} + u_{i2}\bar{u}_{j2} + \cdots + u_{in}\bar{u}_{jn}$$
$$= \delta_{ij}.$$

This equation states that the row vectors of U are mutually orthogonal unit vectors of $\mathscr{V}_n(\mathscr{C})$. Another way of expressing this fact is by the matrix equation

$$U\bar{U}^t = I.$$

This equation implies furthermore that $\bar{U}^t = U^{-1}$, and hence $\bar{U}^t U = I$. Taking complex conjugates, we get $U^t \bar{U} = I$, and therefore the column vectors of U are also mutually orthogonal unit vectors of $\mathscr{V}_n(\mathscr{C})$.

DEFINITION. A matrix U is said to be *unitary* if $U\bar{U}^t = I$.

We have therefore shown above that a matrix of order n is unitary if and only if its row vectors (and therefore its column vectors also) are mutually orthogonal unit vectors of $\mathscr{V}_n(\mathscr{C})$. By reference to Theorem 5.4, the student should now have no difficulty in completing the proof of

THEOREM 9.4. Let $E_1, \cdots, E_n$ be a normal orthogonal basis of $\mathscr{V}_n(\mathscr{C})$, and let U be the matrix of the transformation of coordinates to a new basis $F_1, \cdots, F_n$. A necessary and sufficient condition that the F-basis be normal orthogonal is that U be unitary.

Let X and Y be any two vectors of $\mathscr{V}_n(\mathscr{C})$, and let X' and Y' be their coordinate vectors relative to an arbitrary normal orthogonal basis $E_1, \cdots, E_n$. If U is the matrix of the transformation of coordinates from the basis $(1, 0, \cdots, 0), \cdots, (0, 0, \cdots, 1)$ to the E-basis, we have $X' = XU$, $Y' = YU$, and therefore, since U is unitary by Theorem 9.4,

$$X' \cdot Y' = X'\bar{Y}'^t = XU(\overline{YU})^t = XU\bar{U}^t\bar{Y}^t = X\bar{Y}^t = X \cdot Y.$$

We conclude therefore, as in the real case, that, in defining the inner product of X and Y to be $X\bar{Y}^t$, the coordinate vectors of X and Y relative to any normal orthogonal basis may be used.

Exercise 9.2

1. Prove that, if U is unitary, so also are U^{-1}, $\bar{U}$, and U^t.
2. Prove that the product of two unitary matrices is unitary.
3. Prove that the determinant of a unitary matrix has absolute value 1.
4. Prove that every characteristic root of a unitary matrix has absolute value 1.
5. Prove that every orthogonal matrix is unitary and that a real unitary matrix is orthogonal.

A linear transformation of $\mathscr{V}_n(\mathscr{C})$ whose matrix relative to a normal orthogonal basis is unitary is called a *unitary transformation*. By Theorem 6.8 and Problems 1 and 2, Exercise 9.2, it follows that the matrix of a unitary transformation relative to every normal orthogonal basis is unitary. By methods similar to those used in Theorem 6.10, we can prove

THEOREM 9.5. A linear transformation τ of $\mathscr{V}_n(\mathscr{C})$ preserves inner products if and only if it is unitary.

Preservation of inner products implies, of course, preservation of lengths and orthogonality of vectors. In fact, like the orthogonal transformations of $\mathscr{V}_n(\mathscr{R})$, unitary transformations can be characterized as those linear transformations that preserve the length of every vector of $\mathscr{V}_n(\mathscr{C})$. If τ_1 and τ_2 are transformations having this property, it is clear that the inverses τ_1^{-1}, τ_2^{-1}, and the product $\tau_1\tau_2$ have the same property and are therefore unitary. This also follows by consideration of the matrices of τ_1 and τ_2 and by application of the results of Problems 1 and 2, Exercise 9.2. The unitary transformations of $\mathscr{V}_n(\mathscr{C})$ therefore constitute a group called the *unitary group*. It is the group of all length-preserving linear transformations of $\mathscr{V}_n(\mathscr{C})$.

Exercise 9.3

1. Prove Theorem 9.5.
2. Prove that a linear transformation of $\mathscr{V}_n(\mathscr{C})$ preserves inner products if and only if it preserves lengths.
3. Prove that a unitary transformation of $\mathscr{V}_n(\mathscr{C})$ preserves the r-volume of every r-dimensional parallelepiped.

51. HERMITIAN MATRICES, FORMS, AND TRANSFORMATIONS

A matrix H is said to be *Hermitian* if it is its own adjoint; that is, if $\bar{H}^t = H$. An obvious calculation shows that, if H is Hermitian and U is unitary, then $U^{-1}HU$ is also Hermitian. It follows that, if we define a *Hermitian transformation* to be a linear transformation of $\mathscr{V}_n(\mathscr{C})$ whose matrix relative to a normal orthogonal basis is Hermitian, our definition is independent of the particular normal orthogonal basis chosen. We note also that, if a Hermitian matrix is real, it is symmetric and conversely every real symmetric matrix is Hermitian. We shall see that Hermitian matrices play the same role in the complex theory that symmetric matrices play in the real. For example, we can prove

THEOREM 9.6. If H is a Hermitian matrix, there exists a unitary matrix U such that $U^{-1}HU$ is a diagonal matrix.

The proof of this theorem will be omitted, partly because of its similarity to the proof of Theorem 7.9 and partly because the theorem is a special case of Theorem 9.14 which will be proved in Section 52.

If $H = (h_{ij})$ is Hermitian, it follows from $\bar{H}^t = H$ that $\bar{h}_{ii} = h_{ii}$, $i = 1, 2, \cdots, n$, and the diagonal elements of a Hermitian matrix are therefore real. Hence, if we choose U as in Theorem 9.6, the matrix $U^{-1}HU$ is Hermitian and diagonal and is therefore real. Since its diagonal elements are the characteristic roots of H, we have proved the following

COROLLARY. The characteristic roots of a Hermitian matrix are real.

Theorem 7.8 is included in this corollary as a special case.

Let X vary over $\mathscr{V}_n(\mathscr{C})$, and let $H = (h_{ij})$ be any Hermitian matrix. The function

$$f = XH\bar{X}^t = \sum_{i=1}^{n}\sum_{j=1}^{n} h_{ij}x_i\bar{x}_j$$

is called a *Hermitian form*. It is a function whose domain is $\mathscr{V}_n(\mathscr{C})$ and whose range is a subset of the complex numbers. As a matter of fact, a Hermitian form can take only real values since

$$\overline{XH\bar{X}^t} = \bar{X}H X^t = (X\bar{H}^t\bar{X}^t)^t = (XH\bar{X}^t)^t = XH\bar{X}^t.$$

THEOREM 9.7. Every Hermitian form $XH\bar{X}^t$ is equivalent under a unitary transformation of coordinates in $\mathscr{V}_n(\mathscr{C})$ to the form

(3) $$k_1y_1\bar{y}_1 + k_2y_2\bar{y}_2 + \cdots + k_ny_n\bar{y}_n,$$

where $k_1, k_2, \cdots, k_n$ are the characteristic roots of H.

Proof. Let U be the unitary matrix which diagonalizes H and whose existence is ensured by Theorem 9.6. The transformation of coordinates $Y = XU$, with matrix U, will transform $XH\bar{X}^t$ into $YD\bar{Y}^t$, where $D = U^{-1}HU$ is the diagonal form of H. The result follows.

The fact that the range of values of a Hermitian form is real can also be deduced from Theorem 9.7, since (3) is clearly real and the range of values is not changed by a transformation of coordinates.

A Hermitian matrix H and the corresponding Hermitian form $XH\bar{X}^t$ are said to be *positive definite* or *non-negative definite*, according as the characteristic roots of H are positive or non-negative. If H has both positive and negative characteristic roots, H and XHX^t are said to be *indefinite*. It is clear from the diagonal form (3) that, if H is indefinite, the range of values of XHX^t is the set of all real numbers, whereas, if H is non-negative or positive definite, the range of XHX^t is the non-negative real numbers. Moreover, if H is positive definite, $XHX^t = 0$ implies $X = O$. The term positive semi-definite is also used in the literature to mean the same as non-negative definite.

THEOREM 9.8. A linear transformation τ of $\mathscr{V}_n(\mathscr{C})$ whose matrix (relative to any basis) is positive definite Hermitian is a magnification.

Proof. By Theorem 9.6 the matrix of τ relative to a suitably chosen basis is a diagonal matrix with positive elements in the diagonal. Hence τ is a magnification.

We now prove the analogue of Theorem 7.14 for $\mathscr{V}_n(\mathscr{C})$.

THEOREM 9.9. Every nonsingular matrix A can be written in one and only one way as a product UH, where U is unitary and H is positive definite Hermitian.

Proof. We first note that $\bar{A}^t A$ is Hermitian, and we shall show that it is positive definite. Let $U_1^{-1} \bar{A}^t A U_1 = D$, where U_1 is unitary and D is a diagonal matrix. Since $D = \bar{U}_1^t \bar{A}^t A U_1 = (\overline{AU_1})^t A U_1$, the ith diagonal element of D is equal to $\bar{X}_i X_i^t$, where X_i^t is the ith column vector of AU_1. The diagonal elements of D are therefore non-negative. They cannot be zero since A, and therefore D, is nonsingular. Hence all the characteristic roots of $\bar{A}^t A$ are positive. Now let D_1 be the diagonal matrix whose diagonal elements are the positive square roots of those of D. The required decomposition is obtained exactly as in Theorem 7.14 by putting $H = U_1 D_1 U_1^{-1}$ and $U = AH^{-1}$.

To prove the uniqueness, suppose that $A = UH = U'H'$ are two such decompositions. Then $\bar{A}^t A = H^2 = H'^2$. By Corollary 1, Theorem 7.7, any characteristic root of H^2 has the form k^2, where k is a characteristic root of H. If X is an eigenvector of H^2 corresponding to k^2, then

$$(H^2 - k^2 I)X^t = (H + kI)(H - kI)X^t = 0.$$

Since H is positive definite, $k > 0$ and $H + kI$ is nonsingular, for otherwise H would have a characteristic root $-k$. Hence $(H - kI)X^t = 0$, and X is an eigenvector of H corresponding to k. Now, if V is a unitary

Vector Spaces over the Complex Field

matrix that diagonalizes H', it also diagonalizes H'^2 and therefore H^2. Its columns are therefore unit eigenvectors of H^2 and, by the above argument, also of H. It follows that V also diagonalizes H. However, the diagonal matrices $V^{-1}HV$ and $V^{-1}H'V$ are identical since their squares are identical and all their diagonal elements are positive. Hence $H = H'$, and, since they are nonsingular, $U = U'$.

The above decomposition of a nonsingular matrix A into the product of a unitary matrix and a positive definite Hermitian matrix is called the *polar factorization* of A since, for the case $n = 1$, it reduces to the unique representation of a nonzero complex number $a + bi$ in the polar form $re^{i\theta}$, $r > 0$. The existence of a polar factorization can be proved for an arbitrary square matrix A but, if A is singular, the Hermitian factor is merely non-negative definite and the unitary factor is not uniquely determined. The proof for singular matrices is given in [6]. A consequence of the polar factorization is

THEOREM 9.10. Every nonsingular linear transformation of $\mathscr{V}_n(\mathscr{C})$ is the product of a unitary transformation and a magnification.

The proof is similar to that of Theorem 7.15.

Exercise 9.4

1. Prove that a Hermitian matrix of order n has n mutually orthogonal eigenvectors.

2. Prove that the characteristic polynomial of a Hermitian matrix has real coefficients.

3. Prove that, if A is a skew symmetric real matrix (i.e., $A^t = -A$), there exists a unitary matrix U such that $U^{-1}AU$ is a diagonal matrix with pure imaginary diagonal elements. *Hint:* consider the matrix iA. Deduce that a real skew symmetric matrix of odd order is singular.

4. Prove that every square matrix A can be written in the form $B + iC$, where B and C are Hermitian. *Hint:* Let

$$B = \frac{A + \bar{A}^t}{2}.$$

5. Modify the proof of Theorem 9.9 to show that every nonsingular matrix A can be written as a product H_1U_1, where H_1 is positive definite Hermitian and U_1 is unitary.

52. NORMAL MATRICES AND TRANSFORMATIONS

In this section we shall characterize the class of all matrices that are unitarily similar to diagonal matrices. First we prove a somewhat stronger analogue of Theorem 7.7.

THEOREM 9.11. If A is any square matrix, there exists a unitary matrix U such that $U^{-1}AU$ is a triangular matrix with the characteristic roots of A in the main diagonal.

Proof. The proof is similar to that of Theorem 7.7. Let $k_1, k_2, \cdots, k_n$ be the characteristic roots of A, and let V_1 be a unit eigenvector of A corresponding to k_1. By Theorem 9.2 (d) there exists a unitary matrix V whose first column vector is V_1^t. The first column vector of AV is therefore $AV_1^t = k_1 V_1^t$, and the first column of $V^{-1}AV = \bar{V}^t AV$ is therefore $(k_1, 0, \cdots, 0)^t$. Hence

$$V^{-1}AV = \begin{pmatrix} k_1 & B_1 \\ O & A_1 \end{pmatrix} ,$$

where A_1 is of order $n-1$ and has characteristic roots $k_2, \cdots, k_n$. The theorem is therefore proved if $n = 2$. We complete the proof by induction as before. Assuming the result for matrices of order $n-1$, there exists a unitary $(n-1)$th-order matrix W such that $W^{-1}A_1 W$ is triangular with $k_2, \cdots, k_n$ in the main diagonal. If we let

$$W_1 = \begin{pmatrix} 1 & O \\ O & W \end{pmatrix} ,$$

the matrix $U = VW_1$ is unitary and $U^{-1}AU$ is triangular with diagonal elements $k_1, k_2, \cdots, k_n$.

DEFINITION. A matrix A is said to be *normal* if it commutes with its adjoint, that is, $A\bar{A}^t = \bar{A}^t A$.

THEOREM 9.12. If A is normal and U is unitary, then $U^{-1}AU$ is normal.
Proof.
$$U^{-1}AU(\overline{U^{-1}AU})^t = U^{-1}AU\bar{U}^t\bar{A}^t U = U^{-1}A\bar{A}^t U$$
$$= U^{-1}\bar{A}^t AU = (U^{-1}\bar{A}^t U)(U^{-1}AU) = (\overline{U^{-1}AU})^t(U^{-1}AU).$$

THEOREM 9.13. A triangular matrix is normal if and only if it is diagonal.

Proof. Let

$$A = \begin{pmatrix} a_{11} & a_{12} & \cdots & a_{1n} \\ 0 & a_{22} & \cdots & a_{2n} \\ \cdot & \cdot & \cdot & \cdot \\ 0 & 0 & \cdots & a_{nn} \end{pmatrix} ,$$

and assume that A is normal. The element in the first row and first column of $A\bar{A}^t$ is $a_{11}\bar{a}_{11} + a_{12}\bar{a}_{12} + \cdots + a_{1n}\bar{a}_{1n}$, and the corresponding element of $\bar{A}^tA$ is $a_{11}\bar{a}_{11}$. Hence

$$a_{11}\bar{a}_{11} + a_{12}\bar{a}_{12} + \cdots + a_{1n}\bar{a}_{1n} = a_{11}\bar{a}_{11},$$

and, since each term of the sum on the left is non-negative, we conclude that $a_{12} = a_{13} = \cdots = a_{1n} = 0$. We now have

$$A = \begin{pmatrix} a_{11} & O \\ O & A_1 \end{pmatrix},$$

where A_1 is triangular of order $n-1$ and is also normal since A is. Repetition of the above argument shows that $a_{23} = a_{24} = \cdots = a_{2n} = 0$ and finally that all the elements above the main diagonal are zero. If the nonzero elements of A are below the main diagonal, the proof is similar.

From Theorems 9.11 through 9.13 we now conclude that, if A is normal, there exists a unitary matrix U such that $U^{-1}AU$ is a diagonal matrix. The converse is also true, for a diagonal matrix D is obviously normal, and hence, if $U^{-1}AU = D$, where U is unitary, $A = UDU^{-1}$ and is normal by Theorem 9.12. We have therefore proved

THEOREM 9.14. *Let A be any matrix. A necessary and sufficient condition that there exist a unitary matrix U such that $U^{-1}AU$ is a diagonal matrix is that A be normal.*

In view of Theorem 9.12 we say that a linear transformation of $\mathscr{V}_n(\mathscr{C})$ is normal if its matrix relative to any (and therefore every) normal orthogonal basis is normal. This is the most general transformation whose matrix relative to a suitably chosen normal orthogonal basis is diagonal. It is easily verified that Hermitian and unitary matrices, and therefore the corresponding linear transformations, are normal. The same is therefore true of orthogonal and real symmetric matrices.

Exercise 9.5

1. If A is any matrix and $A = B + iC$, where B and C are Hermitian (see Problem 4, Exercise 9.4), show that A is normal if and only if $BC = CB$.

2. If $A = UH$ is the polar factorization of A, show that A is normal if and only if $UH = HU$.

3. Prove that a matrix of order n is normal if and only if it has n mutually orthogonal eigenvectors.

4. Prove that unitary and Hermitian matrices are normal.

5. Prove that a matrix A is similar to a diagonal matrix if and only if there exists a positive definite Hermitian matrix H such that $H^{-1}AH$ is normal. (See [9].)

53 THE SPECTRAL DECOMPOSITION

A matrix is said to be nondefective if it is similar to a diagonal matrix. The class of nondefective matrices therefore includes all normal matrices and all matrices whose characteristic roots are distinct.

THEOREM 9.15. Let A be any nondefective matrix of order n. Let $k_1, k_2, \cdots, k_s$ be the distinct characteristic roots of A, and suppose that k_i occurs with multiplicity m_i and hence $m_1 + m_2 + \cdots + m_s = n$. Then A may be written in the form

$$(4) \qquad A = k_1 P_1 + k_2 P_2 + \cdots + k_s P_s,$$

where the matrices P_i are of order n and have the following properties:

(a) Each P_i is idempotent, i.e., $P_i^2 = P_i$.
(b) The P_i are "mutually orthogonal," i.e., $P_i P_j = 0$ for $i \neq j$.
(c) $P_1 + P_2 + \cdots + P_s = I$.
(d) Each P_i commutes with A.

Equation 4 is called the *spectral decomposition* of A.

Proof. Since A is nondefective, it is similar to a diagonal matrix. Its diagonal form can clearly be written

$$(5) \qquad P^{-1}AP = k_1 E_1 + k_2 E_2 + \cdots + k_s E_s,$$

where E_i is a diagonal matrix of order n with 1's in those places in the main diagonal in which k_i occurs in $P^{-1}AP$ and 0's elsewhere. It is also clear that $E_i^2 = E_i$ and, since the k_i are distinct, $E_i E_j = 0$ for $i \neq j$, and $E_1 + E_2 + \cdots + E_s = I$.

From equation 5 we see that

$$A = k_1 P E_1 P^{-1} + k_2 P E_2 P^{-1} + \cdots + k_s P E_s P^{-1}.$$

We now let $P_i = P E_i P^{-1}$ ($i = 1, 2, \cdots, s$), and equation 4 follows. Properties (a), (b), and (c) hold for the E_i's and are easily verified for the P_i's. Since E_i commutes with $P^{-1}AP$, it follows that P_i commutes with A, and hence (d) holds.

COROLLARY. Let $f(x)$ be any polynomial. If a nondefective matrix A has spectral decomposition (4), then

$$f(A) = f(k_1)P_1 + f(k_2)P_2 + \cdots + f(k_s)P_s.$$

Proof. Forming successive powers of A from equation 4 and using (a), (b), and (c), we can see that for any integer ≥ 0

$$A^r = k_1^r P_1 + k_2^r P_2 + \cdots + k_s^r P_s,$$

and the result for arbitrary polynomials follows.

Actually the above representation of $f(A)$ can be extended to power series in A but first a short digression on infinite series of matrices is necessary.

Suppose that $A^{(r)} = (a_{ij}^{(r)})$, $r = 1, 2, 3, \cdots$, is any sequence of nth-order matrices. The matrix series

(6) $$\sum_{r=1}^{\infty} A^{(r)}$$

is said to converge if each of the n^2 series

(7) $$\sum_{r=1}^{\infty} a_{ij}^{(r)}$$

converges. If the series (7) converges to a sum s_{ij}, ($i = 1, 2, \cdots, n$; $j = 1, 2, \cdots, n$), then the matrix $S = (s_{ij})$ is called the sum of the matrix series (6). We leave it to the student to prove that, if $\sum_{r=1}^{\infty} A^{(r)}$ converges to the sum S and P is any nth-order matrix, then $\sum_{r=1}^{\infty} PA^{(r)}$ and $\sum_{r=1}^{\infty} A^{(r)}P$ converge to PS and SP, respectively. In particular, if P is nonsingular, $\sum_{r=1}^{\infty} P^{-1}A^{(r)}P$ converges to $P^{-1}SP$.

It can also be shown (see [1] or [5]) that the series

$$I + A + \frac{1}{2!}A^2 + \cdots + \frac{1}{r!}A^r + \cdots$$

converges for every matrix A. For obvious reasons its sum is denoted by e^A. Now suppose that A is nondefective, of order n, and has distinct characteristic roots $k_1, \cdots, k_s$. If (5) is the diagonal form of A, it is clear that

$$P^{-1}A^r P = k_1^r E_1 + k_2^r E_2 + \cdots + k_s^r E_s,$$

for $r \geq 0$, and therefore

$$P^{-1}e^A P = e^{P^{-1}AP} = e^{k_1}E_1 + e^{k_2}E_2 + \cdots + e^{k_s}E_s,$$

whence

$$e^A = e^{k_1}P_1 + e^{k_2}P_2 + \cdots + e^{k_s}P_s.$$

In other words, the corollary to Theorem 9.15 holds for the exponential function as for polynomials.

As an application of the foregoing discussion we prove

THEOREM 9.16. Every unitary matrix U can be written in the form $U = e^{iH}$, where H is Hermitian. Conversely, if H is Hermitian, e^{iH} is unitary.

Proof. By Problem 4, Exercise 9.2, the characteristic roots of a unitary matrix have absolute value 1. Hence, if U is unitary, its distinct characteristic roots may be written in the form $e^{i\lambda_1}, e^{i\lambda_2}, \cdots, e^{i\lambda_s}$ where $\lambda_1, \lambda_2, \cdots, \lambda_s$ are real. Since U is normal, there exists a unitary matrix V such that

$$V^{-1}UV = e^{i\lambda_1}E_1 + e^{i\lambda_2}E_2 + \cdots + e^{i\lambda_s}E_s$$

is the diagonal form of U. Here $E_1, \cdots, E_s$ are the idempotent diagonal matrices used in the proof of Theorem 9.15. It follows that

$$V^{-1}UV = e^{iD},$$

where

$$D = \lambda_1 E_1 + \lambda_2 E_2 + \cdots + \lambda_s E_s.$$

Since D is a diagonal matrix with real elements, it is Hermitian, and therefore

$$U = Ve^{iD}V^{-1} = e^{iVDV^{-1}} = e^{iH},$$

where $H = VDV^{-1}$ is Hermitian since D is Hermitian and V is unitary.

Conversely, if H is Hermitian, $\lambda_1, \cdots, \lambda_s$ are its distinct characteristic roots, and V is the unitary matrix that diagonalizes H, we have

$$V^{-1}HV = \lambda_1 E_1 + \lambda_2 E_2 + \cdots + \lambda_s E_s,$$

where $\lambda_1, \lambda_2, \cdots, \lambda_s$ are real. As before, it follows that

$$V^{-1}e^{iH}V = e^{iV^{-1}HV} = e^{i\lambda_1}E_1 + e^{i\lambda_2}E_2 + \cdots + e^{i\lambda_s}E_s.$$

The right-hand side of this equation is clearly unitary. Hence $V^{-1}e^{iH}V$, and therefore e^{iH}, is also unitary.

Exercise 9.6

1. If the spectral decomposition of A is given by equation 4, show that a matrix commutes with A if and only if it commutes with each P_i. (See [6] p. 128.)

2. If A is nonsingular and has spectral decomposition (4), show that A^{-1} has spectral decomposition

$$A^{-1} = k_1^{-1}P_1 + k_2^{-1}P_2 + \cdots + k_r^{-r}P_r.$$

3. If $k_1, \cdots, k_n$ are the characteristic roots of an arbitrary matrix A of order n, prove that the characteristic roots of e^A are $e^{k_1}, \cdots, e^{k_n}$. (*Hint*: see Theorem 7.7, Corollary 1.) Deduce that for every matrix A, e^A is nonsingular.

4. Prove that, if $AB = BA$, then $e^A e^B = e^{A+B}$.

5. Prove that $(e^A)^{-1} = e^{-A}$.

6. Prove that every proper orthogonal matrix P has the form $P = e^S$, where S is a real skew symmetric matrix (i.e., $S^t = -S$), and that every matrix of this form is proper orthogonal.

7. Calculate e^S for

$$S = \begin{pmatrix} 0 & \theta \\ -\theta & 0 \end{pmatrix}$$

by forming successive powers of S and substituting in the series for e^S.

54. THE REAL CANONICAL FORM OF AN ORTHOGONAL MATRIX

Let P be any orthogonal matrix, and let τ be the linear transformation of $\mathscr{V}_n(\mathscr{C})$ whose matrix is P relative to the normal orthogonal basis $E_1 = (1, 0, \cdots, 0)$, $E_2 = (0, 1, \cdots, 0), \cdots, E_n = (0, 0, \cdots, 1)$. The transformation induced by τ in $\mathscr{V}_n(\mathscr{R})$ also has matrix P relative to the E-basis. It will cause no confusion if we denote this transformation also by τ since it will be clear from the context in which space τ is operating.

By Theorem 7.1 each eigenvector of P^t generates a one-dimensional subspace of $\mathscr{V}_n(\mathscr{C})$ which is left invariant by τ. Since P^t is also orthogonal (and therefore unitary), all its characteristic roots have absolute value 1. The real characteristic roots of P^t are therefore 1 or -1, and the complex ones occur in conjugate pairs of the form $e^{i\theta}$ and $e^{-i\theta}$. Suppose that these characteristic roots are so ordered that the first r of them are equal to 1, the next s are -1, and the last $n-r-s$ are $e^{i\theta_1}, e^{-i\theta_1}, \cdots, e^{i\theta_q}, e^{-i\theta_q}$, where $q = (n-r-s)/2$. We now denote the corresponding unit eigenvectors by

$$X_1, \cdots, X_r, X_{r+1}, \cdots, X_{r+s}, Y_1, \bar{Y}_1, \cdots, Y_q, \bar{Y}_q,$$

since conjugate characteristic roots give rise to conjugate eigenvectors. Since P^t is normal, these eigenvectors can be chosen so that they constitute a normal orthogonal basis of $\mathscr{V}_n(\mathscr{C})$. The effect of τ on this basis is to leave $X_1, \cdots, X_r$ invariant, to map $X_{r+1}, \cdots, X_{r+s}$ on their negatives, and to map Y_j and $\bar{Y}_j$, respectively, onto

$$Y_j^\tau = e^{i\theta} Y_j = (\cos\theta + i\sin\theta) Y_j$$

and

$$\bar{Y}_j^\tau = e^{-i\theta} \bar{Y}_j = (\cos\theta - i\sin\theta) \bar{Y}_j.$$

If we now write

$$Z_j = \frac{Y_j + \bar{Y}_j}{\sqrt{2}},$$

$$Z'_j = \frac{Y_j - \bar{Y}_j}{i\sqrt{2}} \qquad (j = 1, 2, \cdots, q),$$

it is easily verified that Z_j and Z'_j are mutually orthogonal unit vectors of $\mathscr{V}_n(\mathscr{R})$ and hence that

(8) $\qquad X_1, \cdots, X_r, X_{r+1}, \cdots, X_{r+s}, Z_1, Z'_1, \cdots, Z_q, Z'_q$

is a normal orthogonal basis of $\mathscr{V}_n(\mathscr{R})$.

The effect of τ on $X_1, \cdots, X_{r+s}$ has already been noted. The images under τ of Z_j and Z'_j are

$$Z_j^\tau = \frac{Y_j^\tau + \bar{Y}_j^\tau}{\sqrt{2}} = Z_j \cos\theta - Z'_j \sin\theta$$

and

$$Z'^\tau_j = \frac{Y_j^\tau - \bar{Y}_j^\tau}{i\sqrt{2}} = Z_j \sin\theta + Z'_j \cos\theta.$$

Hence the matrix of τ relative to the basis (8) [note that τ is now operating in $\mathscr{V}_n(\mathscr{R})$] is

$$\begin{pmatrix} I_r & & & & & O \\ & -I_s & & & & \\ & & R_1 & & & \\ & & & \cdot & & \\ & & & & \cdot & \\ O & & & & & R_q \end{pmatrix},$$

where I_r and I_s are unit matrices of order r and s and R_j has the form

$$R_j = \begin{pmatrix} \cos\theta_j & -\sin\theta_j \\ \sin\theta_j & \cos\theta_j \end{pmatrix} \qquad (j = 1, 2, \cdots, q).$$

Vector Spaces over the Complex Field

We conclude that $\mathscr{V}_n(\mathscr{R})$ is a direct sum of one- and two-dimensional subspaces which are left invariant by τ. The one-dimensional subspaces are either left pointwise invariant (corresponding to characteristic roots 1) or are reflected in the origin (corresponding to characteristic roots -1). In the jth two-dimensional subspace (generated by Z_j and Z_j') a rotation about the origin through an angle θ_j takes place. The transformation τ is a rotation in $\mathscr{V}_n(\mathscr{R})$ if and only if s is even.

APPENDIX ONE

Abstract Definition of a Vector Space

The algebra of vector spaces is applicable to many situations that have received little or no mention in the foregoing chapters. Especially rich developments are made possible by removing the restriction of a finite dimension. For these reasons it is advantageous to formulate an abstract definition of a vector space over an arbitrary field of scalars that is capable of wider application. To this end we shall give here abstract definitions of the terms, group, field, and vector space, and indicate briefly the role that $\mathscr{V}_n(\mathscr{R})$ and $\mathscr{V}_n(\mathscr{C})$ play in the more general theory.

Abstract algebra deals with sets of undefined *elements* in which certain *rules of composition* or *operations* are postulated. The operations are usually assumed to be binary and everywhere defined in the set under discussion. The set moreover is assumed to be closed under the operations. This means that, if $\mathscr{S}$ is the set of elements, there is an operation whereby any pair of elements a and b of $\mathscr{S}$ may be combined to give an element c of $\mathscr{S}$, usually denoted by ab or $a + b$. Different algebraic systems are obtained by assuming the existence of one or more such operations and by laying down different rules or *postulates* that the operations must satisfy.

The simplest and one of the most important abstract algebraic systems is the *group*. A group is a set $\mathscr{G}$ of abstract elements in which a single operation called group multiplication is defined and satisfies the following postulates:

(a) If a and b are in $\mathscr{G}$, the group product ab is in $\mathscr{G}$.
(b) For all a, b, c in $\mathscr{G}$, $a(bc) = (ab)c$.
(c) There is a unit element e (also called the identity element) in $\mathscr{G}$ such that $ae = ea = a$ for every element a in $\mathscr{G}$.

Abstract Definition of a Vector Space

(*d*) If *a* is in $\mathscr{G}$, there is a uniquely determined element a^{-1} in $\mathscr{G}$, called the inverse of *a*, such that $a^{-1}a = aa^{-1} = e$.

Examples of groups are:

1. The real numbers, group multiplication being *addition*, the unit element 0, and the inverse of *a* being $-a$.
2. The nonzero real numbers, group multiplication being ordinary multiplication.
3. The four complex numbers $1, -1, i, -i$ under multiplication.
4. The numbers $0, 1, 2, \cdots, m - 1$, where *m* is any integer and group multiplication is addition modulo *m*.
5. The numbers $1, 2, \cdots, p - 1$, where *p* is any prime and group multiplication is multiplication modulo *p*.
6. The set of all nonsingular transformations of $\mathscr{V}_n(\mathscr{C})$, the group product being the product of transformations.
7. The set of all *n*th-order orthogonal matrices under matrix multiplication.

If the group multiplication is commutative, i.e., $ab = ba$ for all elements *a*, *b* in $\mathscr{G}$, then $\mathscr{G}$ is called a *commutative* or *Abelian* group.

A *field* is a set $\mathscr{F}$ of elements that is closed under two operations called addition and multiplication. The sum and product of two elements *a* and *b* are denoted by $a + b$ and ab, respectively. The field operations satisfy the three following postulates:

(*a*) The elements of $\mathscr{F}$ form an Abelian group under addition. This is called the additive group of $\mathscr{F}$, and its unit element is denoted by 0.
(*b*) The elements of $\mathscr{F}$ other than 0 form an Abelian group under multiplication, called the multiplicative group of $\mathscr{F}$. Its unit element is denoted by 1.
(*c*) If *a*, *b*, and *c* are elements of $\mathscr{F}$,

$$a(b + c) = ab + ac.$$

Examples of fields are the real numbers, the complex numbers, and the set of all rational functions of *x*. The set of numbers $0, 1, 2, \cdots, p - 1$, where *p* is any prime, form a field with *p* elements if field addition and multiplication are defined to be addition and multiplication modulo *p*.

A *vector space* over a field $\mathscr{F}$ consists of a set $\mathscr{V}$ of undefined elements called vectors. The elements of $\mathscr{F}$ are called scalars. There are two operations: vector addition defined among the elements of $\mathscr{V}$, and multiplication by a scalar defined for any scalar and any vector. These operations satisfy the following postulates:

1. The elements of $\mathscr{V}$ form an Abelian group under addition whose unit element O is called the zero vector. The inverse of V is denoted by $-V$.
2. (a) If $V \in \mathscr{V}$ and $k \in \mathscr{F}$, the scalar product kV is an element of $\mathscr{V}$.
   (b) If $k \in \mathscr{F}$ and $V_1, V_2 \in \mathscr{V}$, $k(V_1 + V_2) = kV_1 + kV_2$.
   (c) If $k_1, k_2 \in \mathscr{F}$ and $V \in \mathscr{V}$, $(k_1 + k_2)V = k_1V + k_2V$ and $k_1(k_2V) = (k_1k_2)V$.
   (d) If 1 is the unit element of $\mathscr{F}$, $1V = V$ for all V in $\mathscr{V}$.

An obvious example of a vector space over $\mathscr{F}$ is the set $\mathscr{V}_n(\mathscr{F})$ of all n-tuples $(x_1, x_2, \cdots, x_n)$, where $x_i \in \mathscr{F}$ and vector addition and scalar multiplication are defined as in $\mathscr{V}_n(\mathscr{R})$ and $\mathscr{V}_n(\mathscr{C})$. Other examples are:

1. The set of all real solutions of the linear differential equation

$$\frac{d^n y}{dx^n} + a_1(x)\frac{d^{n-1}y}{dx^{n-1}} + \cdots + a_n(x)y = 0$$

   is a vector space over the real field.
2. The set of all continuous (or all bounded, or all integrable) functions of x defined on the interval $a \leq x \leq b$ is a vector space over the real field.
3. The set of all "infinite vectors" of the form $(x_1, x_2, \cdots)$, where $x_i \in \mathscr{F}$, $i = 1, 2, \cdots$, and addition and scalar multiplication are defined as in $\mathscr{V}_n(\mathscr{F})$.

If $\mathscr{V}$ is a vector space over $\mathscr{F}$, the vectors $V_1, V_2, \cdots, V_r$ are said to be linearly dependent if there exist elements $a_1, a_2, \cdots, a_r$ of $\mathscr{F}$, not all 0, such that

$$a_1V_1 + a_2V_2 + \cdots + a_rV_r = O.$$

Otherwise $V_1, V_2, \cdots, V_r$ are linearly independent. The space $\mathscr{V}$ is said to have dimension n if it contains n linearly independent vectors, but any $n + 1$ vectors of $\mathscr{V}$ are linearly dependent. If for every integer n, $\mathscr{V}$ contains n linearly independent vectors, then $\mathscr{V}$ is said to be of infinite dimension. Examples 2 and 3 above are spaces of infinite dimension.

Two vector spaces $\mathscr{V}$ and $\mathscr{W}$ over the same field $\mathscr{F}$ are said to be *isomorphic* if there exists a one-to-one mapping of the vectors of $\mathscr{V}$ onto those of $\mathscr{W}$ such that
   (a) If V is mapped on W and $k \in \mathscr{F}$, kV is mapped on kW.
   (b) If V_1 is mapped on W_1 and V_2 on W_2, then $V_1 + V_2$ is mapped on $W_1 + W_2$.

Isomorphic spaces have the same algebraic structure and can differ only in the nature of their elements. From the abstract point of view they may be considered to be identical. It is not difficult to show that two

Abstract Definition of a Vector Space 189

finite-dimensional vector spaces over $\mathscr{F}$ are isomorphic if and only if they have the same dimension. The procedure is to show, just as in Section 7, that, if the space $\mathscr{V}$ over $\mathscr{F}$ has dimension n, any n linearly independent vectors $E_1, \cdots, E_n$ constitute a basis of $\mathscr{V}$. Every vector V then has a unique representation in the form

$$x_1 E_1 + x_2 E_2 + \cdots + x_n E_n.$$

The one-to-one correspondence

$$V \leftrightarrow (x_1, x_2, \cdots, x_n)$$

which maps V on its coordinate vector relative to the E-basis is then an isomorphism between $\mathscr{V}$ and the space $\mathscr{V}_n(\mathscr{F})$ of all coordinate vectors. Hence every n-dimensional space over $\mathscr{F}$ is isomorphic to $\mathscr{V}_n(\mathscr{F})$. In this sense, therefore, restriction of the definition of a vector space to spaces of n-tuples over the field $\mathscr{F}$ involves no loss of generality in the finite-dimensional case. Restriction of the field of scalars to be the real or complex numbers is more drastic since these fields, although the most useful from the point of view of applications, are not typical of the general mathematical situation. The complex numbers have the special property of *algebraic closure*. A field $\mathscr{F}$ is algebraically closed if every polynomial equation with coefficients in $\mathscr{F}$ has a root in $\mathscr{F}$. This property ensures for example that the characteristic roots of a linear transformation in $\mathscr{V}_n(\mathscr{F})$ lie in the field of scalars. This is important, for example, in the discussion in Section 38 where the real case had to be treated differently from the complex. The lack of algebraic closure of $\mathscr{R}$ is not too serious, however, since we never hesitate to call in the complex numbers when we need them. For a discussion of the general problem of similarity of matrices over an arbitrary field the reader is referred to [14].

A vector space $\mathscr{E}$ over the field of real numbers is said to be *Euclidean* if, for any two vectors X and Y of $\mathscr{E}$, a real valued inner product $X \cdot Y$ is defined which satisfies the four postulates IM1 through 4 listed in Section 2. These postulates are sufficient to establish for an abstract Euclidean space of finite dimension the entire theory of orthogonality. Most of this can also be done for certain infinite-dimensional spaces.

If $\mathscr{E}_n$ is a Euclidean vector space of dimension n, it is easy to determine all possible inner products that can be defined in $\mathscr{E}_n$. Suppose $E_1, \cdots, E_n$ is a basis of $\mathscr{E}_n$, and suppose $E_i \cdot E_j = h_{ij}$. Since inner multiplication is commutative, $H = (h_{ij})$ is a real symmetric matrix. Now let $V = \sum_{i=1}^{n} x_i E_i$ and $W = \sum_{i=1}^{n} y_i E_i$ be any two vectors of $\mathscr{E}_n$, and let X and Y be their coordinate vectors relative to the E-basis. By IM3 and IM4

$$V \cdot W = \sum_{i=1}^{n} \sum_{j=1}^{n} x_i y_j E_i \cdot E_j = \sum_{i=1}^{n} \sum_{j=1}^{n} h_{ij} x_i y_j = XHY^t.$$

Hence $V \cdot W$ is a symmetric bilinear form in the coordinates of V and W. Moreover, since $V \cdot V = XHX^t \geq 0$ and is 0 only if $X = O$, it follows that H is positive definite. Hence the most general inner product that can be defined in $\mathscr{E}_n$ is that defined by XHY^t, where H is a positive definite symmetric matrix. Conversely an arbitrary bilinear form with positive definite symmetric matrix defines an inner product that satisfies IM1 through 4.

A vector space over the field of complex numbers is called a *unitary space* if there is defined for any two vectors X and Y a complex-valued inner product $X \cdot Y$ which satisfies the four postulates IMC1 through 4 listed in Section 49. Just as we did above for Euclidean spaces, we can show that the most general inner product in a unitary space of finite dimension n is

$$V \cdot W = \sum_{i=1}^{n} \sum_{j=1}^{n} h_{ij} x_i \bar{y}_j = XH\bar{Y}^t,$$

where X and Y are coordinate vectors of V and W relative to a basis $E_1, \cdots, E_n$ and $H = (E_i \cdot E_j)$ is a positive definite Hermitian matrix. Conversely every positive definite Hermitian bilinear form $XH\bar{Y}^t$ defines an inner product in the unitary space.

Infinite-dimensional unitary spaces are of great interest. They include Hilbert space, an example of which is the space of all complex-valued functions $f(x)$ defined on the real unit interval $0 \leq x \leq 1$, and whose squares are Lebesgue-integrable on that interval. The inner product is defined by

$$f(x) \cdot g(x) = \int_0^1 f(x) \overline{g(x)} \, dx.$$

The extension of the concept of a normal orthogonal basis to this space is intimately connected with the expansion of functions in Fourier series. For information about Hilbert space the reader is referred to [10].

APPENDIX TWO

Three-Dimensional Analytic Geometry

A2.1. THE REAL NUMBERS

We shall not require a detailed knowledge of the real number system, but the student should be familiar with the following definitions and basic facts even though we may not make specific use of all of them. The natural numbers 1, 2, 3, 4, $\cdots$ are called the positive integers. These together with 0 and their negatives $-1, -2, -3, -4, \cdots$ make up the *integers*. A *rational number* is a number that can be expressed as a quotient p/q of two integers, of which the denominator q is not 0. The sum, difference, product, and quotient of two rational numbers is again a rational number. The decimal representation of a rational number is either a terminating or repeating decimal, and conversely every terminating or repeating decimal represents a rational number. The *real numbers* are usually defined by convergent sequences of rational numbers. For our purposes a real number may be thought of as any finite or infinite decimal, although such a decimal must, of course, have only a finite number of digits to the left of the decimal point. Every real number that is not rational is said to be *irrational*. The irrational numbers are, therefore, those infinite decimals that do not repeat. Rational approximations to a real number r may be found to any required degree of accuracy by terminating the decimal representation of r after the nth place for sufficiently large values of n. A complete theory of infinite decimals requires a discussion of convergence of infinite sequences. Since for our purposes an intuitive knowledge of the real numbers is sufficient, we shall make no attempt to fill in the logical foundations. The student who is interested in this subject is referred to [2] or [7]. The student should also be familiar with complex numbers, that is, numbers of the form $a + bi$, where a and

b are real and $i^2 = -1$. An elementary discussion of these will be found in [3]; a more sophisticated treatment, in [2].

A2.2. COORDINATES ON A LINE

By a line we shall always mean a straight line extending to infinity in both directions. If A and B are two points on a line, we shall use the notation AB both for the line segment between A and B and for the length of this segment. It will be clear from the context which of these two meanings is intended. Let l be any straight line. We shall set up a one-to-one correspondence between the real numbers and the points of l in the following way. Corresponding to the real number 0, we choose an arbitrary point P_0 on l, and, corresponding to the real number 1, we choose any point P_1 on l different from P_0. If the line l is horizontal as in Figure A2.1, it is usual to choose P_1 to the right of P_0. The point P_0 is called the *origin* and will often be designated by the letter O rather than by P_0. The point P_1 is called the unit point. For reasons that will appear later, points of l on the same side of the origin as P_1 will be said to lie on the *positive* side of the origin, and points on the side of the origin

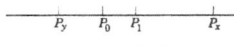

Figure A2.1

opposite to P_1 are said to lie on the *negative* side of the origin. Using P_0P_1 as a unit of distance, the point P_x corresponding to a positive real number x is chosen as the point on the positive side of the origin whose distance from the origin is x. The point P_x corresponding to a negative real number x is chosen on the negative side at a distance $|x|$ from the origin. Conversely, to any point P on l there corresponds a uniquely determined real number x such that $P_0P = |x|$ and x is positive or negative according as P lies on the positive or negative side of the origin. Designating by P_x the point corresponding to the real number x, we say that x is the *coordinate* of P_x relative to the origin P_0 and the unit point P_1. If x and y are any two real numbers (positive, negative, or zero), the length of the segment P_xP_y is $|x - y|$, the unit of length being P_0P_1. The student should check this statement for positive and negative values of x and y.

The reader who has more than an intuitive knowledge of the real number system will realize that the foregoing discussion ignores several difficulties and conceals a basic assumption concerning the real numbers

Three-Dimensional Analytic Geometry

and the set of all points on a line. For example, given a real number x, how is the corresponding point P_x to be found? If x is rational or a square root of a rational number, the point P_x can be located by a simple geometric construction with ruler and compass. However, for many irrational numbers, for example, $\sqrt[3]{2}$ and π, it can be proved that no such ruler-and-compass construction exists. In general, no method of locating the point P_x can be given. Conversely, given a point P on l, how is its coordinate to be found? Actual measurement is, of course, only approximate. Can we be sure that a real number exists that will "measure" every line segment P_0P in terms of a given unit P_0P_1? What we are actually assuming is that there exists a one-to-one correspondence $x \leftrightarrow P_x$ between the real numbers x and the points P_x of l such that

(a) P_x is to the left of P_y if and only if $x < y$.
(b) The segment P_xP_y has length $|x - y|$ in terms of the unit P_0P_1.

Exercise A2.1

1. Given a line l, origin P_0, and unit point P_1, describe a ruler-and-compass construction that will locate the point P_x (a) when x is a rational number p/q; (b) when $x = \sqrt{2}$; (c) when $x = \sqrt{n}$, n being a positive integer.

2. Ruler-and-compass constructions are discussed in [3]. Look up this reference, and find out for what values of x the point P_x can be thus located.

A2.3. COORDINATES IN THE PLANE

In a plane p choose any two straight lines l_1 and l_2 that intersect in O. Choose O as origin on each line, and choose unit points P_1 on l_1 and Q_1

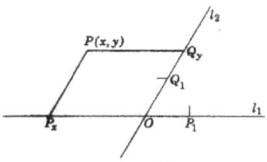

Figure A2.2

on l_2, making $OP_1 = OQ_1$. As described above, we set up coordinates on each of the lines l_1 and l_2. For any real number t we let P_t be the point on l_1 and Q_t the point on l_2 corresponding to t. Now let P be any point in the plane p. As illustrated in Figure A2.2, we draw lines through

P parallel to l_1 and l_2, and let these cut l_1 in P_x and l_2 in Q_y. The two real numbers (x, y), named in that order, uniquely determine the point P. They are called the *coordinates* of P relative to the *coordinate axes* l_1 and l_2 and the unit points P_1 and Q_1.

Just as the conventional procedure is to choose the unit point to the right of the origin on a horizontal line, so there are conventions about the choice of the coordinate axes and unit points in the plane. Normally l_1 is drawn horizontal and P_1 chosen to the right of O. Now let the segment OP_1 be rotated about O, like the spoke of a wheel, in the counter-clockwise direction, and let Q_1 be the point in which P_1 first meets l_2. Then Q_1 is chosen as the unit point on l_2. Coordinates on l_1 are designated by x, and for this reason l_1 will be called the *x-axis*. Similarly coordinates on l_2 (the *y-axis*) are designated by y. For most purposes the coordinate axes are assumed to intersect at right angles, in which case the x-axis is horizontal and the y-axis vertical. The unit point on l_1 is then to the right of O, and the unit point on l_2 above O. The coordinates of a point are then called rectangular Cartesian coordinates.

A2.4. COORDINATES IN SPACE

Let l_1, l_2, and l_3 be any three straight lines that intersect in O and do not lie in one plane. We assign coordinates to the points of each line, choosing O as origin in each case. Let the unit points P_1 on l_1, Q_1 on l_2, and R_1 on l_3 be chosen so that $OP_1 = OQ_1 = OR_1$. Now let P be any point in space. Referring to Figure A2.3, a plane through P, parallel

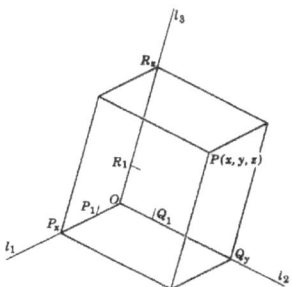

Figure A2.3

to the plane of l_2 and l_3 will cut l_1 in a uniquely determined point P_x whose coordinate is x. Similarly let the plane through P parallel to the plane of l_1 and l_3 cut l_2 in Q_y with coordinate y, and the plane through P parallel to the plane of l_1 and l_2 cut l_3 in R_z with coordinate z. The three numbers (x, y, z) are uniquely determined by P and are called Cartesian coordinates of P. Conversely, any three real numbers (x, y, z), named in that order, determine a unique point P whose coordinates they are.

Coordinates on l_1, l_2, and l_3 are designated by the letters x, y, and z, respectively. For this reason these lines will be referred to as the x-, y-, and z-axes, respectively, and collectively as the coordinate axes. For most purposes the coordinate axes will be chosen to be three mutually perpendicular lines, and the corresponding coordinates will then be called rectangular Cartesian coordinates. Unless otherwise stated, we shall assume coordinates to be rectangular. We shall designate the origin by O, the positive sides of the x-, y-, and z-axes by OX, OY, and OZ. The plane containing the x- and y-axes will be called the xy-plane, with similar definitions for the yz-plane and the xz-plane. We shall adopt the following conventions for choice of unit points on the coordinate axes. The unit points P_1, Q_1, and R_1 on the x-, y-, and z-axes will be chosen so that an observer standing at R_1 and looking at the xy-plane will find that P_1 and Q_1 have been chosen in accordance with the convention laid down above for the case of two coordinate axes in the plane. Such a set of coordinate axes is called a *right-hand system*.

A2.5. THE DISTANCE FORMULA

Let $A_1(x_1, y_1, z_1)$ and $A_2(x_2, y_2, z_2)$ be any two points in space. Through A_1 and A_2 pass planes parallel to the yz-plane. These planes are perpendicular to the x-axis and will cut it in points P_{x_1} and P_{x_2}, whose linear coordinates are x_1 and x_2. It follows that the perpendicular distance between these parallel planes is $|x_1 - x_2|$. Similarly, if planes parallel to the xz-plane are passed through A_1 and A_2, they will cut the y-axis in Q_{y_1} and Q_{y_2}, whose distance apart is $|y_1 - y_2|$, and planes through A_1 and A_2 parallel to the xy-plane will cut the z-axis in R_{z_1} and R_{z_2} whose distance apart is $|z_1 - z_2|$. From Figure A2.4 it is seen that these six planes contain the six faces of a rectangular parallelepiped (i.e., box). The segment A_1A_2 is a diagonal of this box, and the lengths of the three edges A_1B, A_1C, and A_1D are, respectively, $|x_1 - x_2|$, $|y_1 - y_2|$, and $|z_1 - z_2|$. By drawing the diagonal A_1A_2 of the box and the diagonal A_1E of the face A_1BEC, we see, from the right-angled triangle A_1A_2E, that

$$A_1A_2^2 = A_1E^2 + EA_2^2,$$

and, from the right triangle A_1CE,

$$A_1E^2 = A_1C^2 + CE^2.$$

From these two equations we deduce that

$$A_1A_2^2 = A_1C^2 + CE^2 + EA_2^2,$$

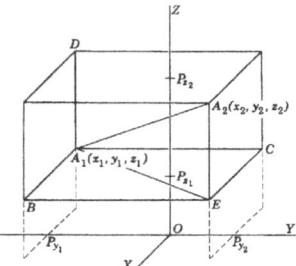

Figure A2.4

or that the square of the diagonal of a rectangular box is the sum of the squares of its three edges. Since it was shown above that the edges of this box are the absolute values of the differences in coordinates of A_1 and A_2, we have

(1) $\qquad A_1A_2 = \sqrt{(x_1 - x_2)^2 + (y_1 - y_2)^2 + (z_1 - z_2)^2}.$

A2.6. DIRECTION. PARAMETRIC EQUATIONS OF A LINE

When a coordinate system has been chosen a straight line is determined if we know either the coordinates of two points on it or the coordinates of one point and the direction of the line. The direction of a line must, of course, be defined with reference to the coordinate axes. Moreover, since on any line there are two directions, we shall introduce the concept of half-line or *ray*, which is that part of a line starting at a fixed *initial point* and extending to infinity in one direction only. The *direction* of a ray is the direction along the ray away from its initial point. Consider a ray whose initial point is the origin. Its direction is determined by the three angles that the ray makes with the positive sides of the x-, y-, and z-axes. Let these three angles be α, β, and γ. They are called the direction angles of the ray. Since $0 \leq \alpha \leq 180°$, and similar inequalities hold for β and γ, these angles are uniquely determined by their cosines. Consider the case, illustrated in Figure A2.5, in which α, β, and γ are

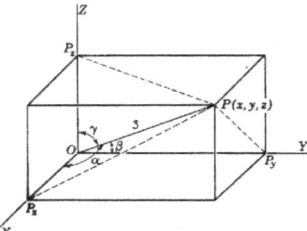

Figure A2.5

all acute. If $P(x, y, z)$ is any point other than O on the ray and if $OP = s$, we see from the right triangles OPP_x, OPP_y, and OPP_z that

(2) $$\cos \alpha = x/s, \quad \cos \beta = y/s, \quad \cos \gamma = z/s.$$

If α, β, γ are not all acute, equations 2 still hold since the changes in sign of $\cos \alpha$, $\cos \beta$, and $\cos \gamma$ when the angles become obtuse are taken care of by corresponding changes in sign of x, y, and z. Similarly, if $x = 0$, P is on the yz-plane and $\alpha = 90°$, and the other cases in which P is on a coordinate plane or axis may be similarly verified. These three cosines are called the *direction cosines* of the ray OP and are designated respectively, by the letters λ, μ, and ν. Since $x^2 + y^2 + z^2 = s^2$, we see that

(3) $$\lambda^2 + \mu^2 + \nu^2 = \frac{x^2 + y^2 + z^2}{s^2} = 1.$$

Now consider any ray AB whose initial point is A. A uniquely determined ray OP with initial point O can be drawn parallel to, and similarly directed to, AB. The direction cosines of AB are then defined to be the same as those of OP. Hence we may conclude that

The sum of the squares of the direction cosines of any ray is equal to 1.

A line l through the origin is made up of two oppositely directed rays with O as initial point. If α, β, γ are the direction angles of one of these rays, then $180° - \alpha$, $180° - \beta$, $180° - \gamma$ are the direction angles of the other. Hence, if λ, μ, ν are the direction cosines of either one of these two rays, $-\lambda$, $-\mu$, $-\nu$ are the direction cosines of the other. Either λ, μ, ν or $-\lambda$, $-\mu$, $-\nu$ are, therefore, said to be a set of direction cosines of the line l. Direction cosines of an arbitrary line are defined to be the same as the direction cosines of a parallel line through the origin. It follows that two lines are parallel if and only if they have the same direction cosines. The fact that there are two sets of direction cosines for any line will not embarrass us. Either set, together with a point on the line, will uniquely determine the line. For a point and a set of direction cosines uniquely determines a ray emanating from the point. This ray in turn uniquely determines the line.

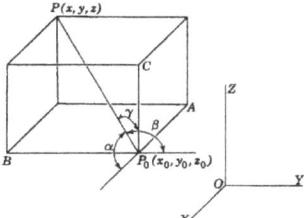

Figure A2.6

Now consider any ray P_0P with initial point $P_0(x_0, y_0, z_0)$ and direction cosines λ, μ, ν. Let $P(x, y, z)$ be any other point on the ray. As was done in Figure A2.4, a rectangular box with faces parallel to the coordinate planes can be constructed with P_0P as diagonal. The lengths of the edges P_0A, P_0B, and P_0C of this box are, respectively, $|x - x_0|$, $|y - y_0|$, and $|z - z_0|$. If, from initial point P_0, rays are drawn parallel to and in the same sense as OX, OY, and OZ, the direction angles of P_0P are the angles α, β, and γ, shown in Figure A2.6, which P_0P makes with these rays. It is clear that $\alpha > 90°$ if and only if $x - x_0 < 0$, and hence from the right triangle P_0PA we see that in all cases (since $x - x_0 = 0$, if $\alpha = 90°$)

$$\lambda = \cos \alpha = \frac{x - x_0}{s},$$

Three-Dimensional Analytic Geometry

where s is the length of P_0P. Similarly the student should deduce that

$$\mu = \cos \beta = \frac{y - y_0}{s}$$

$$\nu = \cos \gamma = \frac{z - z_0}{s}.$$

Solving these equations for the coordinates (x, y, z) of an arbitrary point on the ray, we have

(4)
$$\begin{aligned} x &= x_0 + \lambda s, \\ y &= y_0 + \mu s, \qquad 0 \leq s < \infty. \\ z &= z_0 + \nu s, \end{aligned}$$

These are parametric equations of the ray. The point (x, y, z) traverses the ray as s runs through the non-negative real numbers. Now consider equations 4 in which $-\infty < s \leq 0$. These can obviously be rewritten, letting $s' = -s$, in the form

$$\begin{aligned} x &= x_0 - \lambda s', \\ y &= y_0 - \mu s', \qquad 0 \leq s' < \infty \\ z &= z_0 - \nu s', \end{aligned}$$

and are, therefore, parametric equations of the oppositely directed ray, that is, the ray with initial point (x_0, y_0, z_0) and direction cosines $-\lambda, -\mu, -\nu$. It follows that the equations

(5)
$$\begin{aligned} x &= x_0 + \lambda s, \\ y &= y_0 + \mu s, \qquad -\infty < s < \infty \\ z &= z_0 + \nu s \end{aligned}$$

are parametric equations of the line through (x_0, y_0, z_0) with direction cosines λ, μ, ν, and that positive values of the parameter s give points of the line on one side of (x_0, y_0, z_0), whereas negative values of s give points of the line on the other side of (x_0, y_0, z_0).

Now suppose that (x_1, y_1, z_1) is the point on the line corresponding to $s = s_1 \neq 0$. We then have from equations 5

$$\frac{x_1 - x_0}{\lambda} = \frac{y_1 - y_0}{\mu} = \frac{z_1 - z_0}{\nu} = s_1.$$

and hence the direction cosines λ, μ, ν of the line are proportional to the differences $x_1 - x_0$, $y_1 - y_0$, $z_1 - z_0$. Since the points (x_1, y_1, z_1) and (x_0, y_0, z_0) are arbitrary, we may state:

The direction cosines of a line are proportional to the differences of corresponding coordinates of any two distinct points on the line.

If λ, μ, ν are direction cosines of a line, any three numbers proportional to λ, μ, ν are called *direction numbers* of the line. Any three real numbers a, b, c not all of which are zero are direction numbers of a line in space, for example, the line joining the origin to the point $P(a, b, c)$. Direction cosines of OP are then obtained from equations 2 by dividing a, b, and c by $OP = \sqrt{a^2 + b^2 + c^2}$. In general, *direction cosines of a line may be obtained from any set of direction numbers by dividing by the square root of the sum of their squares.*

Example. Find direction cosines of the line joining $P(2, 6, -3)$ and $Q(0, 4, 1)$.

Solution. The direction cosines of PQ are proportional to the coordinate differences $2 - 0$, $6 - 4$, $-3 - 1$; hence to $2, 2, -4$ or $1, 1, -2$. Direction cosines of PQ are therefore $1/\sqrt{6}, 1/\sqrt{6}, -2/\sqrt{6}$.

A2.7. THE ANGLE BETWEEN TWO RAYS

If two distinct or coincident rays r_1 and r_2 have the same initial point, we shall understand by the angle between them the least non-negative angle θ measured from one ray to the other. It follows that $0 \leq \theta \leq 180°$. If the two rays r_1 and r_2 do not have the same initial point, we draw from the initial point of r_1 a ray r_3 parallel to and in the same sense as r_2. The angle between r_1 and r_2 is then defined to be the angle between r_1 and r_3. Now suppose that r_1 and r_2 have direction cosines λ_1, μ_1, ν_1 and λ_2, μ_2, ν_2, respectively. Let r_1' and r_2' be the rays with the same direction cosines as r_1 and r_2, but with the origin as initial point. Then, from equations 5, r_1' and r_2' have parametric equations

(6)
$$x = \lambda_1 s, \qquad x = \lambda_2 s,$$
$$y = \mu_1 s, \quad \text{and} \quad y = \mu_2 s,$$
$$z = \nu_1 s, \qquad z = \nu_2 s,$$

and the angle θ between r_1' and r_2' is the same as that between r_1 and r_2. Putting $s = 1$ in equations 6, we find that the points at unit distance from the origin on r_1' and r_2', respectively, are $A_1(\lambda_1, \mu_1, \nu_1)$ and $A_2(\lambda_2, \mu_2, \nu_2)$.

Applying the law of cosines to the triangle OA_1A_2, we find, since $OA_1 = OA_2 = 1$,

$$\cos\theta = \frac{OA_1^2 + OA_2^2 - A_1A_2^2}{2(OA_1)(OA_2)}$$

$$= \frac{2 - (\lambda_1 - \lambda_2)^2 - (\mu_1 - \mu_2)^2 - (\nu_1 - \nu_2)^2}{2},$$

by (1),

$$= \lambda_1\lambda_2 + \mu_1\mu_2 + \nu_1\nu_2,$$

by (3). More generally, if $P(x_1, y_1, z_1)$ and $P_2(x_2, y_2, z_2)$ are any two points on the rays r_1' and r_2', respectively, and if $OP_1 = s_1$ and $OP_2 = s_2$, we have from equations 6 that

$$\lambda_1 = \frac{x_1}{s_1}, \mu_1 = \frac{y_1}{s_1}, \nu_1 = \frac{z_1}{s_1} \quad \text{and} \quad \lambda_2 = \frac{x_2}{s_2}, \mu_2 = \frac{y_2}{s_2}, \nu_2 = \frac{z_2}{s_2}.$$

If θ is the angle between OP_1 and OP_2, we have therefore

$$\cos\theta = \lambda_1\lambda_2 + \mu_1\mu_2 + \nu_1\nu_2 = \frac{x_1x_2 + y_1y_2 + z_1z_2}{s_1s_2}.$$

Two rays are said to be perpendicular or *orthogonal* if the angle between them is a right angle. In this case $\cos\theta = 0$, and we may therefore state:

Two rays with direction cosines λ_1, μ_1, ν_1 and λ_2, μ_2, ν_2 are orthogonal if and only if

(7) $$\lambda_1\lambda_2 + \mu_1\mu_2 + \nu_1\nu_2 = 0.$$

Since in equation 7, λ_1, μ_1, ν_1 or λ_2, μ_2, ν_2 can be replaced by their negatives without destroying the equality, it follows that (7) is also a necessary and sufficient condition for the orthogonality of two *lines* whose direction cosines are λ_1, μ_1, ν_1 and λ_2, μ_2, ν_2. It is also a useful fact that in (7) the direction cosines of either line may be replaced by any three numbers proportional to them, since (7) is equivalent to

$$(k_1\lambda_1)(k_2\lambda_2) + (k_1\lambda_1)(k_2\mu_2) + (k_1\nu_1)(k_2\nu_2) = 0,$$

provided that neither k_1 nor k_2 is zero. Hence, if two lines l_1 and l_2 have direction numbers a_1, b_1, c_1 and a_2, b_2, c_2, a necessary and sufficient condition that l_1 and l_2 be orthogonal is that

(7') $$a_1a_2 + b_1b_2 + c_1c_2 = 0.$$

It should be emphasized that orthogonal lines need not intersect each other. However, a line drawn parallel to one of these lines and intersecting the other will intersect it at right angles.

Exercise A2.2

1. Given the points $P(2, -1, 6)$ and $Q(1, 3, -2)$, find:
   (a) Direction cosines of the line PQ.
   (b) Parametric equations of PQ.
   (c) The length of PQ.
   (d) The coordinates of the mid-point of PQ.
   (e) The points in which the line through P and Q pierces each of the three coordinate planes.

2. Find the three interior angles of the triangle whose vertices are $(2, 0, -1)$, $(3, 1, -1)$, $(2, -3, 4)$.

3. Find direction cosines of:
   (a) The x-axis, y-axis, and z-axis.
   (b) The line in the xy-plane that bisects the first and third quadrants.
   (c) The ray that makes equal angles with OX, OY, and OZ.
   (d) A line perpendicular to each of two lines whose direction numbers are $2, -1, 6$ and $2, 2, 1$.

A2.8. THE EQUATION OF A PLANE

Let p be any plane, and let r be a ray, with initial point at the origin, perpendicular to p and intersecting p. The ray r is thus uniquely determined by p unless p passes through the origin, in which case either one of the two possible rays may be chosen for r. Now let $P_1(x_1, y_1, z_1)$ be any fixed point and $P(x, y, z)$ an arbitrary point on the plane p. The line joining P to P_1 has direction numbers $x - x_1$, $y - y_1$, and $z - z_1$ and is orthogonal to r since it lies in the plane p. Hence, if λ, μ, and ν are the direction cosines of r, we have by (7')

(8) $$\lambda(x - x_1) + \mu(y - y_1) + \nu(z - z_1) = 0.$$

This equation is satisfied by the coordinates (x, y, z) of every point on the plane p. Moreover, if (x, y, z) is not on p, then PP_1 is not orthogonal to r and (8) is not satisfied. Equation 8 is therefore called the equation of the plane p. We may also write (8) in the form

(9) $$\lambda x + \mu y + \nu z = \lambda x_1 + \mu y_1 + \nu z_1.$$

It is clear that the constant term $\lambda x_1 + \mu y_1 + \nu z_1$ is independent of the particular fixed point (x_1, y_1, z_1) chosen on p, since equation 9 says in effect that the expression $\lambda x + \mu y + \nu z$ takes the same value at all points of the plane. If we choose $P_1(x_1, y_1, z_1)$ to be the point in which the ray r cuts the plane p, and if we let d be the length of OP_1, then we have, from equations 2,

$$\lambda d = x_1,$$
$$\mu d = y_1,$$
$$\nu d = z_1,$$

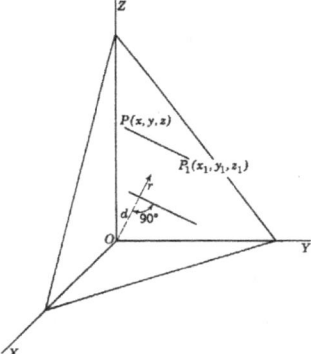

Figure A2.7

whence $\lambda x_1 + \mu y_1 + \nu z_1 = (\lambda^2 + \mu^2 + \nu^2)d = d$. Hence equation 8 becomes

(10) $$\lambda x + \mu y + \nu z = d.$$

Here λ, μ, and ν are the direction cosines of the ray drawn from the origin perpendicular to and intersecting the plane, and d is the length of the perpendicular from the origin to the plane.

Now consider any first-degree equation in x, y, z, namely,

(11) $$ax + by + cz = k,$$

where not all the coefficients a, b, and c are zero. The ray drawn from the origin through the point (a, b, c) has direction cosines

$$\frac{a}{\sqrt{a^2 + b^2 + c^2}}, \quad \frac{b}{\sqrt{a^2 + b^2 + c^2}}, \quad \frac{c}{\sqrt{a^2 + b^2 + c^2}}.$$

Hence, if we write equation 11 in the form

$$\frac{ax}{\pm \sqrt{a^2 + b^2 + c^2}} + \frac{bx}{\pm \sqrt{a^2 + b^2 + c^2}} + \frac{cx}{\pm \sqrt{a^2 + b^2 + c^2}}$$
$$= \frac{k}{\pm \sqrt{a^2 + b^2 + c^2}}$$

and compare with (10), we see that (11) is the equation of a plane that is perpendicular to a line with direction cosines

$$\lambda = \frac{a}{\pm \sqrt{a^2 + b^2 + c^2}}, \quad \mu = \frac{b}{\pm \sqrt{a^2 + b^2 + c^2}}, \quad \nu = \frac{c}{\pm \sqrt{a^2 + b^2 + c^2}}$$

and whose perpendicular distance from the origin is $\left| \dfrac{k}{\sqrt{a^2 + b^2 + c^2}} \right|$.

If $k \neq 0$ and the sign of the radical is chosen to agree with that of k, then λ, μ, ν are the direction cosines of the ray drawn from the origin, perpendicular to the plane, and intersecting the plane. A line or ray drawn perpendicular to a plane is called a *normal* to the plane. We may now state the important result:

Every first-degree equation in x, y, z of the form $ax + by + cz = k$ is the equation of a plane and any normal to this plane has direction cosines proportional to a, b, and c.

The following examples illustrate methods of finding the equation of a plane when geometric conditions sufficient to fix its position are given.

Example 1. Find the equation of a plane that passes through the point $(1, -2, 6)$ and is perpendicular to a line with direction numbers $-3, 2, 4$.

Since in equation 8 the direction cosines of the normal may be replaced by any set of direction numbers, the required equation is

$$-3(x - 1) + 2(y + 2) + 4(z - 6) = 0$$

or

$$3x - 2y - 4z + 17 = 0.$$

Example 2. Find the equation of the plane that passes through the three points $P(2, 1, 0)$, $Q(-1, 5, 2)$, and $R(3, 1, 4)$.

Any normal to the plane is orthogonal to every line in the plane and in particular therefore to PQ and to QR. Let λ, μ, ν be direction cosines of such a normal. Direction numbers of PQ are $2 + 1, 1 - 5$, and $0 - 2$ or $3, -4, -2$. Similarly those of QR are $-2, 2, -1$. Hence by (7') we have

$$3\lambda - 4\mu - 2\nu = 0,$$
$$-2\lambda + 2\mu - \nu = 0.$$

Solving for λ and μ in terms of ν gives

$$\frac{\lambda}{8} = \frac{\mu}{7} = \frac{\nu}{-2}.$$

Any normal to the required plane therefore has direction numbers $8, 7, -2$, and, since the plane passes through $P(2, 1, 0)$ its equation may be written

$$8(x - 2) + 7(y - 1) - 2(z - 0) = 0$$

or

$$8x + 7y - 2z - 23 = 0.$$

A2.9. PARALLEL AND PERPENDICULAR PLANES. THE LINE OF INTERSECTION OF TWO PLANES

Using the above result, we can now find the conditions that two planes are parallel or perpendicular to each other. Let

(12)
$$a_1 x + b_1 y + c_1 z = d_1,$$
$$a_2 x + b_2 y + c_2 z = d_2$$

be the equation of two planes. Any normals to these two planes have direction numbers a_1, b_1, c_1 and a_2, b_2, c_2, respectively.

Since the planes are parallel if and only if their normals are parallel, it follows that *the planes* (12) *are parallel if and only if*

(13)
$$\frac{a_1}{a_2} = \frac{b_1}{b_2} = \frac{c_1}{c_2}.$$

(The condition expressed in equations 13 must be interpreted so that, if one or more of the coefficients of one of equations 12 are zero, then the corresponding coefficients in the other equation are also zero.) Similarly the planes are perpendicular if and only if their normals are perpendicular. Hence *a necessary and sufficient condition that the planes* (12) *be perpendicular is that*

$$a_1 a_2 + b_1 b_2 + c_1 c_2 = 0.$$

Any two planes that are not parallel intersect in a straight line. Hence all points whose coordinates satisfy both equations 12 must be on the line of intersection of these planes. Conversely, every point on this line must have coordinates that satisfy both equations 12. For this reason equations 12 are called equations of this line. In general a line in space is defined by the equations of any two distinct planes both of which contain the line, and these are called *equations of the line*. It is clear that equations of a line are by no means uniquely determined. In fact, if k_1 and k_2 are any two real numbers, not both 0, the equation

(14) $\quad k_1(a_1x + b_1y + c_1z - d_1) + k_2(a_2x + b_2y + c_2z - d_2) = 0$

is satisfied by all values of x, y, z that satisfy both equations 12. Since (14) is linear in x, y, z, it is the equation of a plane. Hence for real values of k_1 and k_2, not both 0, equation 14 represents a plane through the line of intersection of the planes (12).

Direction cosines of the line (12) may be easily found. Suppose that they are λ, μ, ν. Since the line lies in each of the two defining planes, it is perpendicular to the normal to each of these planes. Since these normals have direction numbers, a_1, b_1, c_1 and a_2, b_2, c_2, we have

$$a_1\lambda + b_1\mu + c_1\nu = 0,$$
$$a_2\lambda + b_2\mu + c_2\nu = 0.$$

From these equations we find

$$\frac{\lambda}{\begin{vmatrix} b_1 & c_1 \\ b_2 & c_2 \end{vmatrix}} = \frac{\mu}{-\begin{vmatrix} a_1 & c_1 \\ a_2 & c_2 \end{vmatrix}} = \frac{\nu}{\begin{vmatrix} a_1 & b_1 \\ a_2 & b_2 \end{vmatrix}}$$

or

$$\frac{\lambda}{b_1c_2 - b_2c_1} = \frac{\mu}{a_2c_1 - a_1c_2} = \frac{\nu}{a_1b_2 - a_2b_1}.$$

These equations, together with (3), determine direction cosines of the line. By finding the coordinates of a fixed point on the line, parametric equations for it can be found from (5).

Example. Find direction cosines and parametric equations of the line

(15) $\quad\quad\quad x - 2y + 4z = 6,$
$\quad\quad\quad\quad\quad 2x + y - 3z = 8.$

Let λ, μ, ν be direction cosines of this line. Then

$$\lambda - 2\mu + 4\nu = 0,$$
$$2\lambda + \mu - 3\nu = 0,$$

Three-Dimensional Analytic Geometry

whence $\lambda/2 = \mu/11 = \nu/5$. Hence 2, 11, 5 are direction numbers of the line, and direction cosines are

$$\lambda = \frac{2}{\sqrt{150}}, \quad \mu = \frac{11}{\sqrt{150}}, \quad \text{and} \quad \nu = \frac{5}{\sqrt{150}}.$$

To find parametric equations of the line we require the coordinates of a fixed point on it. Putting $z = 0$ in (15) and solving the resulting equations, we find $x = \frac{22}{5}$ and $y = -\frac{4}{5}$. Hence $(\frac{22}{5}, -\frac{4}{5}, 0)$ is a point on the line, and from (5) we can write parametric equations of the line (15) in the form

$$x = \frac{22}{5} + \frac{2}{\sqrt{150}} s$$

$$y = -\frac{4}{5} + \frac{11}{\sqrt{150}} s$$

$$z = \frac{5}{\sqrt{150}} s.$$

By changing the parameter to $t = \dfrac{s}{\sqrt{150}}$, we could use instead the equations

(16)
$$x = \tfrac{22}{5} + 2t,$$
$$y = -\tfrac{4}{5} + 11t,$$
$$z = 5t.$$

The student should check that, if the values of x, y, and z given in (16) are substituted in (15), these equations are identically satisfied, for all values of t. This shows that the line (16) lies in each of the planes (15) and is, indeed, therefore, their line of intersection.

Exercise A2.3

1. Find the equation of the plane through the point $P(2, -1, 7)$ that is perpendicular to the line joining P to the origin.

2. Find the equation of the plane through the point $(1, -6, 3)$ that is parallel to the plane $2x - 7y + z = 14$.

3. Find the equation of the plane through the point $(4, 2, -1)$ that is perpendicular to the line joining $(1, -2, 6)$ and $(4, 3, 7)$.

4. Find the equation of the plane through the three points $(1, -3, 4)$, $(2, 0, 1)$, and $(3, 1, 0)$.

5. Find the equation of the plane that passes through $(1, -5, 0)$ and is perpendicular to the line of intersection of the planes $x - y + 2z = 0$ and $4x + y - z = 2$.

6. Find the point in which the line joining the points $(2, -1, 0)$ and $(2, 0, 4)$ pierces the plane $2x - y + z = 6$.

7. Prove that the line of intersection of the planes $2x - y - z = 1$ and $x + 3y + 2z = 3$ lies in the plane $7x + 7y + 4z = 11$.

8. Find the point common to the planes $x + y - z = 6$, $2x - 3y + z = 4$, and $3x + 2y - 2z = 12$.

9. Find the equation of the plane that contains the point $(2, 0, -1)$ and the line of intersection of the planes $2x - y + z = 6$ and $x - 3y + 2z = 9$.

A2.10. SURFACES IN THREE-DIMENSIONAL SPACE

It will be assumed that the reader is familiar with the relationship in two-dimensional analytic geometry between an equation and its graph. The graph, or locus, of an equation $f(x, y) = 0$ is the set of all points, and only those points, whose coordinates (x, y) satisfy this equation. The graph may take various forms, some of which are illustrated by the following examples:

(a) The equation $x^2 + y^2 = -7$ is not satisfied by any pair of real numbers x and y. There are therefore no points on its graph.

(b) The graph of the equation $(x - 1)^2 + (y - 2)^2 = 0$ is the single point $(1, 2)$.

(c) The graph of the equation

$$(x^2 + y^2)[x^2 + (y - 1)^2] = 0$$

consists of the two isolated points $(0, 0)$ and $(0, 1)$.

(d) The graph of the equation $x^2 + \sin^2 y = 0$ consists of an infinite number of isolated points $(0, n\pi)$, where n is any integer.

(e) The graph of the equation $(x - 1)^2 + (y + 2)^2 = 16$ is a circle with center at $(1, -2)$ and radius 4.

Although the first four of these examples illustrate interesting possibilities, the student is no doubt aware that they are not typical of the graphs studied in analytic geometry. Normally the locus of an equation $f(x, y) = 0$ is a *curve* as is the case in (e). The term curve here is to be interpreted broadly and includes a straight line as a special case. The equations in (a) through (d) may be given a somewhat different interpretation by considering "points" whose coordinates are complex numbers. From this point of view, by analogy with the equation $x^2 + y^2 = 7$, the equation $x^2 + y^2 = -7$ is said to represent an *imaginary circle* since

Three-Dimensional Analytic Geometry

all the "points" whose coordinates satisfy the equation are "imaginary points" with one or more imaginary coordinates. Similarly the equation $(x - 1)^2 + (y - 2)^2 = 0$ in (b) may be written

$$y - 2 = \pm \sqrt{-1}\,(x - 1)$$

and is said to represent two imaginary lines intersecting in the real point (1, 2). Although terminology of this kind is useful and revealing and may be used from time to time, the graph of an equation for most purposes consists only of the real points whose coordinates satisfy it.

A similar situation exists in space. The set of all points (x, y, z) whose coordinates satisfy an equation of the form

(17) $$F(x, y, z) = 0$$

constitutes the *locus* or *graph* of this equation. This locus is normally, though subject to exceptions analogous to those in examples (a) through (d), a *surface*. This surface may be a plane, as we have seen is the case when equation 17 is linear in x, y, z, or it may be a curved surface. The term surface, like the term curve in plane geometry, will not be defined precisely. The following examples will illustrate its meaning:

Example 1. From equation 1 it is known that the square of the distance from the origin to the point (x, y, z) is $x^2 + y^2 + z^2$. Hence all the points whose coordinates satisfy the equation

$$x^2 + y^2 + z^2 = 16$$

are 4 units distant from (0, 0, 0) and therefore lie on the surface of a sphere with center at the origin and radius 4. Conversely the coordinates of every point on the sphere satisfy this equation. Similarly,

$$(x - 2)^2 + (y - 3)^2 + (z + 6)^2 = 20$$

is the equation of a sphere with center at (2, 3, − 6) and radius $\sqrt{20}$.

Example 2. What is the locus *in three-dimensional space* of the equation $9x^2 + 4y^2 = 36$?

It is known that *in the plane* this equation represents an ellipse with semi-axes 2 and 3. Hence, if we draw in the xy-plane the ellipse whose equation in the plane is $9x^2 + 4y^2 = 36$, the points $(x, y, 0)$ on this ellipse certainly satisfy the equation and form part of the three-dimensional graph. However, a point (x, y, k) satisfies the equation if and only if $(x, y, 0)$ does. Moreover the points (x, y, k), k arbitrary, are precisely the points of the line through $(x, y, 0)$ perpendicular to the xy-plane. The surface, a segment of which is shown in Figure A2.8, is therefore composed of all lines perpendicular to the xy-plane that intersect that plane in a point of the ellipse. Such a surface is called an *elliptic cylinder*.

Example 3. A point moves so that the line joining it to the origin makes a constant angle θ with the positive side of the z-axis. Find the equation of the locus.

Let $P(x, y, z)$ be any point on the locus. Draw a line PQ perpendicular to the z-axis to meet either OZ or OZ' in Q (see Figure A2.9). The

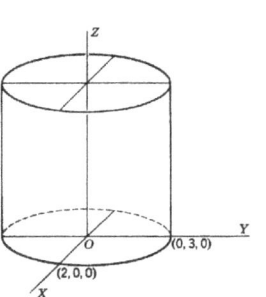

Figure A2.8

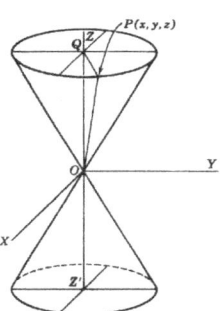

Figure A2.9

coordinates of Q are then $(0, 0, z)$ and $PQ^2 = x^2 + y^2$. Since $OQ^2 = z^2$ and since P is on the locus if and only if $QP^2 = OQ^2 \tan^2 \theta$, we have

$$x^2 + y^2 = z^2 \tan^2 \theta,$$

which is the equation of the locus. The locus is a cone with vertex at the origin and axis of symmetry along the z-axis. The surface extends to infinity both above and below the xy-plane. It may be thought of as the surface generated by revolving the line OP about the z-axis.

A2.11. CURVES IN SPACE

Two surfaces that intersect each other ordinarily intersect in a curve, although there are exceptional cases in which the intersection is itself a surface or in which it reduces to one or more isolated points. We shall assume, unless otherwise stated, that the two surfaces considered intersect in a curve. This curve may lie in a plane in which case it is called a *plane curve*, or it may not in which case it is called a *space curve* or skew curve. If the equations of the two intersecting surfaces are

(18)
$$f_1(x, y, z) = 0,$$
$$f_2(x, y, z) = 0,$$

then the curve of intersection consists of the points (x, y, z), whose coordinates satisfy *both* equations 18. A curve in space therefore is defined by two equations in x, y, z, and the curve consists of those points whose coordinates are simultaneous solutions of these equations. The equations of a curve are not uniquely determined by the curve. For example, if k_1 and k_2 are real numbers different from zero, either of equations 18 can be replaced by

$$k_1 f_1(x, y, z) + k_2 f_2(x, y, z) = 0,$$

and the resulting pair of equations will have the same simultaneous solutions as (18) and therefore represent the same curve. If equations 18 are linear, our discussion reduces to that of Section A2.9 in which the equations of a line are defined as the equations of any two planes that intersect in the given line. It should be mentioned also that it is usually possible to define a curve in space by parametric equations analogous to (5), in which the coordinates of an arbitrary point on the curve are given as functions of a parameter.

A2.12. CYLINDERS

Many surfaces can be generated by a line that moves according to some prescribed condition. Of particular importance among the surfaces so generated are the cylinders, an example of which was discussed in Section A2.10.

A *cylinder* is a surface generated by a straight line that moves so that it remains perpendicular to a fixed plane and its point of intersection with this plane generates a curve.

A cylinder is therefore composed of straight lines perpendicular to a fixed plane and cutting this plane in a curve. These lines are called *generating lines*, *generators*, or sometimes *elements* of the cylinder. All generating lines of a cylinder are parallel to each other since they are perpendicular to a fixed plane. All plane sections perpendicular to the generating lines are congruent. Suppose that the generating lines of a cylinder are parallel to the z-axis and the equations of the curve in which the cylinder cuts the xy-plane are

$$f(x, y) = 0,$$
$$z = 0.$$

Then, for the reason described in Example 2, Section A2.10, the equation of this cylinder is $f(x, y) = 0$. In general, *if one of the three variables x, y, and z does not occur in the equation of a surface, then the surface is a cylinder with generators parallel to the axis of the missing variable.* For example, the equation $4x^2 + z^2 = 16$ is an elliptic cylinder with generators parallel to the y-axis and cutting the plane $y = k$ in the ellipse

$$4x^2 + z^2 = 16,$$
$$y = k.$$

Now consider any space curve C defined by equations 18. If through every point of this curve a line is drawn perpendicular to the xy-plane, these lines are generators of a cylinder that cuts the xy-plane in another curve C', called the (orthogonal) projection of C on the xy-plane. The cylinder is called the projecting cylinder of C onto the xy-plane.

Example 1. Find the equations of the projections of the curve

(19)
$$x^2 + y^2 + z^2 = 25,$$
$$x + 2y - z = 0$$

onto the coordinate planes.

The first equation is that of a sphere with center at the origin and radius 5. The second is the equation of a plane through the origin. The given curve is therefore a circle with center at the origin and lying in the plane $x + 2y - z = 0$.

Solving the second equation of (19) for z and substituting in the first, we find

(20)
$$x^2 + y^2 + (x + 2y)^2 = 25 \quad \text{or}$$
$$2x^2 + 4xy + 5y^2 = 25.$$

Equation 20 is satisfied by all simultaneous solutions of (19) and hence is the equation of a surface containing the curve (19). Since every simultaneous solution of

(21)
$$2x^2 + 4xy + 5y^2 = 25,$$
$$x + 2y - z = 0$$

also satisfies $x^2 + y^2 + z^2 = 25$, it follows that the equations 21 define the same curve as equations 19. Since (20) does not contain z, it is the equation of a cylinder with generators parallel to the z-axis and is therefore the projecting cylinder of the curve (19) onto the xy-plane. This cylinder cuts the xy-plane in the curve

$$2x^2 + 4xy + 5y^2 = 25,$$
$$z = 0,$$

which is therefore the projection of the given curve onto the xy-plane.

By similar methods the projections on the yz- and xz-planes are found to be

$$5y^2 - 4yz + 2z^2 = 25,$$
$$x = 0$$

and

$$5x^2 - 2xz + 5z^2 = 100,$$
$$y = 0.$$

All three projections are ellipses.

This example illustrates the general principle that the equation of the projecting cylinder of a curve with generators parallel to the x-, y-, or z-axis can normally be found by eliminating x, y, or z, respectively, from the equations of the curve.

Example 2. Find the equations of the projection on the xy-plane of the line

$$x - 2y + z = 6,$$
$$3x - y + 2z = 2.$$

The equation

$$2(x - 2y + z - 6) - (3x - y + 2z - 2) = 0,$$

which reduces to

$$x + 3y + 10 = 0,$$

is the equation of a plane through the given line parallel to the z-axis. Hence the projection in the xy-plane is the line

$$x + 3y + 10 = 0,$$
$$z = 0.$$

A2.13. CONES

A surface is called a *cone* if it is generated by straight lines all of which intersect a given plane curve and pass through a fixed point not in the plane of this curve.

The generating lines are called *generators* or *elements* of the cone, and the fixed point through which all the generators pass is called the *vertex* of the cone.

Example. Find the equation of the cone with vertex at $V(0, 0, 4)$ that cuts the xy-plane in the ellipse

$$4x^2 + y^2 = 4,$$
$$z = 0.$$

Let $P(x, y, z)$ be any point on the cone other than the vertex, and let $Q(a, b, 0)$ be the point in which VP cuts the xy-plane. Since V, P, and Q are collinear, direction numbers of QV and PV are proportional, and hence

$$\frac{a-0}{x-0} = \frac{b-0}{y-0} = \frac{0-4}{z-4}$$

Solving for a and b, we find that the coordinates of Q are

$$\left(\frac{4x}{4-z}, \frac{4y}{4-z}, 0\right).$$

It is clear that the point P is on the cone if and only if Q is on the ellipse $4x^2 + y^2 = 4$, $z = 0$; that is, if and only if

$$4\left[\frac{16x^2}{(4-z)^2}\right] + \frac{16y^2}{(4-z)^2} = 4$$

or

$$16x^2 + 4y^2 - (z-4)^2 = 0.$$

Since the coordinates of V also satisfy this equation, it is the equation of the cone.

A2.14. SURFACES OF REVOLUTION

Let $x = f(y)$, $z = 0$ be equations of a curve in the xy-plane. If this curve is rotated about the y-axis, it generates a surface with circular

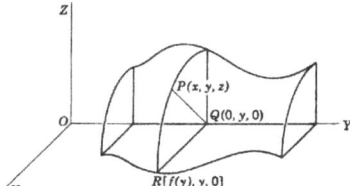

Figure A2.10

cross sections in any plane perpendicular to the y-axis. Such a surface is called a *surface of revolution*. To find its equation we refer to Figure A2.10 and note that, if $P(x, y, z)$ is on the surface, a plane through P perpendicular to the y-axis cuts the y-axis in $Q(0, y, 0)$ and cuts the given curve in $R[f(y), y, 0]$. Since $QP^2 = QR^2$, we have

$$x^2 + z^2 = [f(y)]^2,$$

which is the equation of the surface.

A2.15. THE QUADRIC SURFACES

The simplest curved surfaces are the quadric surfaces, whose equations are of the second degree in x, y, and z. In addition to the quadric cylinders and cones, examples of which were given in Examples 2 and 3, Section A2.9, there are five nondegenerate real quadric surfaces: the ellipsoid, two hyperboloids, and two paraboloids. We shall describe these surfaces by deriving their principal features from certain standard second-degree equations. A general classification of all surfaces that can be represented by a second-degree equation in x, y, and z will be found in Chapter 8. The method used here will be to sketch the surface from its equation by finding the curves in which it cuts the coordinate planes and other planes parallel to them. The curve of intersection of a surface and a plane is sometimes called the *trace* of the surface in the plane.

The ellipsoid. Consider the surface whose equation is

$$\frac{x^2}{a^2} + \frac{y^2}{b^2} + \frac{z^2}{c^2} = 1.$$

In this equation, and in what follows, a, b, and c are assumed to be positive. Directly from its equation we can draw several conclusions about this surface. It is symmetric with respect to each of the three coordinate planes since, if (x, y, z) is on the surface, so also are $(-x, y, z)$, $(x, -y, z)$, and $(x, y, -z)$. If (x, y, z) is a point on the surface, then $|x| \leq a$, $|y| \leq b$, and $|z| \leq c$. The surface is therefore bounded, lying inside the box bounded by the planes $x = \pm a$, $y = \pm b$, and $z = \pm c$ and touching the faces of this box at the points $(\pm a, 0, 0), (0, \pm b, 0)$, and $(0, 0, \pm c)$. The trace of the surface in the xy-plane is the ellipse

$$\frac{x^2}{a^2} + \frac{y^2}{b^2} = 1,$$

$$z = 0.$$

If $|k| < c$, the trace in the plane $z = k$ is the ellipse

$$\frac{x^2}{a^2\left(1 - \frac{k^2}{c^2}\right)} + \frac{y^2}{b^2\left(1 - \frac{k^2}{c^2}\right)} = 1,$$
$$z = k,$$

whose semi-axes decrease as $|k|$ increases. Similarly the traces in the planes $x = k$, where $|k| < a$, and $y = k$, where $|k| < b$, are also ellipses. It is now possible to sketch the surface a segment of which is shown in Figure A2.11. This surface is called an *ellipsoid*. If two of the numbers

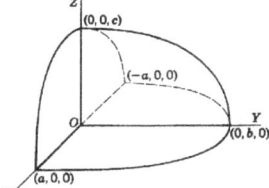

Figure A2.11. Ellipsoid.

a, b, and c are equal, the ellipsoid is an *ellipsoid of revolution*, the surface generated by revolving an ellipse about its major or minor axis. If $a = b = c$, the surface is a sphere with center at the origin.

The hyperboloid of one sheet. Consider the surface whose equation is

$$\frac{x^2}{a^2} + \frac{y^2}{b^2} - \frac{z^2}{c^2} = 1.$$

Its trace in the xy-plane is the ellipse

$$\frac{x^2}{a^2} + \frac{y^2}{b^2} = 1,$$
$$z = 0$$

and, in the plane $z = k$, the ellipse

$$\frac{x^2}{a^2\left(1 + \frac{k^2}{c^2}\right)} + \frac{y^2}{b^2\left(1 + \frac{k^2}{c^2}\right)} = 1,$$
$$z = k.$$

Three-Dimensional Analytic Geometry

The semi-axes of the ellipse in which the surface cuts the plane $z = k$ therefore increase as $|k|$ increases.

The trace in the xz-plane is the hyperbola

$$\frac{x^2}{a^2} - \frac{z^2}{c^2} = 1,$$

$$y = 0,$$

and the trace in the yz-plane is the hyperbola

$$\frac{y^2}{b^2} - \frac{z^2}{c^2} = 1,$$

$$x = 0.$$

A sketch of the surface, which is called a *hyperoboloid of one sheet*, is shown in Figure A2.12. If $a = b$, the trace in any plane $z = k$ is a circle

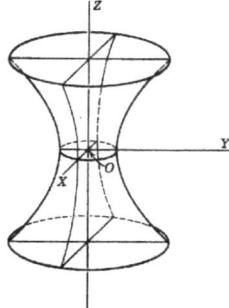

Figure A2.12. Hyperboloid of one sheet.

and the surface is a hyperboloid of revolution generated by revolving a hyperbola about its conjugate axis.

The hyperboloid of two sheets. The surface whose equation is

$$-\frac{x^2}{a^2} + \frac{y^2}{b^2} - \frac{z^2}{c^2} = 1$$

does not cut the xz-plane nor indeed any plane $y = k$, where $|k| < b$. However, if $|k| > b$, its trace in the plane $y = k$ is the ellipse

$$\frac{x^2}{a^2} + \frac{z^2}{b^2} = \frac{k^2}{b^2} - 1,$$

$$y = k.$$

In the xy- and yz-planes the traces are, respectively, the hyperbolas

$$\frac{x^2}{a^2} - \frac{y^2}{b^2} = -1, \quad z = 0$$

and
$$\frac{y^2}{b^2} - \frac{z^2}{c^2} = 1, \quad x = 0.$$

The surface is symmetrical with respect to each of the coordinate planes and consists of two parts or *sheets*, one in region $y \geq b$ and one in the region $y \leq -b$. A sketch of the surface, a *hyperboloid of two sheets*, is shown in Figure A2.13. If $a = b$, the surface has circular cross sections

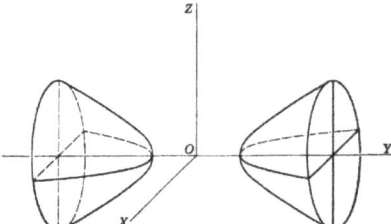

Figure A2.13. Hyperboloid of two sheets.

in planes perpendicular to the y-axis. It is then a hyperboloid of revolution generated by revolving a hyperbola about its transverse axis.

The elliptic paraboloid. The equation

$$\frac{x^2}{a^2} + \frac{y^2}{b^2} = cz$$

represents an *elliptic paraboloid* a sketch of which for the case $c > 0$ is shown in Figure A2.14. If $k > 0$, its trace in the plane $z = k$ is the ellipse

$$\frac{x^2}{cka^2} + \frac{y^2}{ckb^2} = 1,$$

$$z = k.$$

The surface does not extend below the xy-plane. Its traces in the xz- and yz-planes are the parabolas

$$x^2 = a^2cz, \quad y = 0$$

and

$$y^2 = b^2cz, \quad x = 0.$$

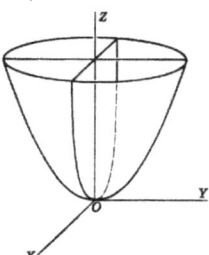

Figure A2.14. Elliptic paraboloid.

The surface is symmetric with respect to the xz- and yz-planes. If $a = b$, it is a paraboloid of revolution.

The hyperbolic paraboloid. The surface whose equation is

$$\frac{y^2}{b^2} - \frac{x^2}{a^2} = cz$$

is called a *hyperbolic paraboloid*. Its trace in the xz-plane is the parabola $x^2 = -a^2cz$, $y = 0$, which opens downward if $c > 0$ and has the z-axis as axis of symmetry. Its trace in the yz-plane is the parabola $y^2 = b^2cz$, $x = 0$, which opens upward if $c > 0$ and has the z-axis as axis of symmetry. Its trace in the plane $z = k$, $k \neq 0$, is the hyperbola

$$\frac{x^2}{a^2ck} - \frac{y^2}{b^2ck} = -1,$$

$$z = k.$$

If $k > 0$, the transverse axis of the hyperbola is parallel to the y-axis, but, if $k < 0$, its transverse axis is parallel to the x-axis. The trace in the plane $z = 0$ is the two straight lines

$$\frac{x^2}{a^2} - \frac{y^2}{b^2} = 0,$$

$$z = 0.$$

A sketch of this surface for the case $c > 0$ is shown in Figure A2.15.

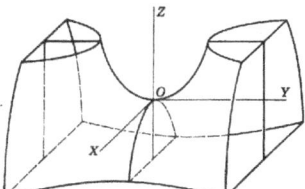

Figure A2.15. Hyperbolic paraboloid.

A2.16. TRANSLATION OF AXES

Let OX, OY, OZ be the positive rays of the x-, y-, and z-axes of a rectangular Cartesian coordinate system. Through an arbitrary point O' with coordinates (x_0, y_0, z_0), draw rays $O'X'$, $O'Y'$, and $O'Z'$ parallel to and in the same sense as OX, OY, and OZ, respectively. It is clear that $O'X'$, $O'Y'$, and $O'Z'$ can be used as the positive rays of a new set of coordinate axes with origin at O'. If a point P has coordinates (x, y, z) relative to the x-, y-, and z-axes and coordinates (x', y', z') with respect to the x'-, y'-, and z'-axes, the student may verify by reference to Figure A2.16 that

(22)
$$x = x_0 + x',$$
$$y = y_0 + y',$$
$$z = z_0 + z'.$$

The change of coordinate axes from OX, OY, OZ to OX', OY', OZ' is called a *translation* of axes. Equations 22 are called the equations of the transformation of coordinates effected by this translation.

Three-Dimensional Analytic Geometry

If $f(x, y, z) = 0$ is the equation of a surface relative to the x-, y-, z-axes, substitution from equations 22 gives

$$f(x' + x_0, y' + y_0, z' + z_0) = 0$$

as the equation of the *same* surface relative to the x'-, y'-, z'-axes.

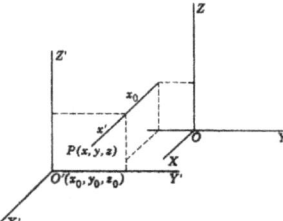

Figure A2.16

Example 1. Consider the surface whose equation is

(23) $$\frac{(x-1)^2}{4} + \frac{(y+2)^2}{9} + \frac{(z-6)^2}{5} = 1.$$

A translation of axes to the new origin $(1, -2, 6)$ is effected by the transformation

$$x = 1 + x',$$
$$y = -2 + y',$$
$$z = 6 + z'.$$

Equation 23 then becomes

$$\frac{x'^2}{4} + \frac{y'^2}{9} + \frac{z'^2}{5} = 1,$$

which is the equation of an ellipsoid.

We conclude that equation 23 represents an ellipsoid with center at $(1, -2, 6)$.

Example 2. What is the surface whose equation is

$$2x^2 + y^2 - 3z^2 + 4x - 2y + 5z = 6?$$

222 **Linear Algebra for Undergraduates**

Completing squares, this equation may be written

$$2(x^2 + 2x + 1) + (y^2 - 2y + 1) - 3(z^2 - \tfrac{2}{3}z + \tfrac{1}{9}) = 6 + 2 + 1 - \tfrac{1}{3}$$

or

$$2(x + 1)^2 + (y - 1)^2 - 3(z - \tfrac{1}{3})^2 = \tfrac{26}{3}.$$

A translation of axes to the new origin $(-1, 1, \tfrac{1}{3})$ reduces this equation to

$$2x'^2 + y'^2 - 3z'^2 = \tfrac{26}{3},$$

which is the equation of a hyperboloid of one sheet whose axis of symmetry is the z'-axis.

Exercise A2.4

1. Find equations of the following surfaces:

    (a) The paraboloid generated by revolving the parabola

    $$z^2 = 4y,$$
    $$x = 0$$

    about the y-axis.

    (b) The cone generated by revolving the line

    $$y = 2x,$$
    $$z = 0$$

    about the x-axis.

    (c) The cylinder with generators parallel to the y-axis that cuts the xz-plane in a circle with radius 3 and center at $(2, 0, -1)$.

2. Identify each of the following surfaces:

    (a) $x^2 - y^2 - z^2 = 1$.
    (b) $x^2 - y^2 - z^2 = 0$.
    (c) $x^2 - y^2 - z^2 = -1$.
    (d) $y^2 - 4z = 0$.
    (e) $3x^2 + 4y^2 = 24$.
    (f) $y^2 - 4z^2 = 0$.
    (g) $y^2 - x^2 = 4z$.

3. Find equations of the projections onto the xy-plane of the curves of intersection of the following pairs of surfaces. Identify the surfaces, and draw figures.

    (a) $y^2 + z^2 = 4$, $x + y + z = 2$.
    (b) $4x^2 + y^2 = 4z^2$, $x^2 + y^2 = 4z$.

4. By suitable translation of axes reduce each of the following equations to a form similar to one of the standard forms discussed in Section A2.14. Hence identify the surfaces represented by each of these equations:

(a) $x^2 + y^2 + z^2 - 2x + y + 6z = 20$.
(b) $x^2 + y^2 - 5x + 2y - 8z + 2 = 0$.
(c) $2x^2 - y^2 + 4z^2 + 6x - 2y + z = 12$.

5. Describe the loci in three-dimensional space that are defined by the following equations:

(a) $x^2 + y^2 + z^2 = -4$.
(b) $(x-1)^2 + (y+2)^2 + (z-5)^2 = 0$.
(c) $x^2 + (y-1)^2 = 0$.
(d) $[x^2 + y^2 + (z-1)^2][(x-1)^2 + (z-2)^2] = 0$.
(e) $(x + y + z - 2)^2 + (x^2 + 4y^2 - 4)^2 = 0$.
(f) $x^2 + y^2 + \sin^2 z = 0$.
(g) $x^2 + \sin^2 z = 0$.

References

1. Bellman, Richard, *Stability Theory of Differential Equations*, McGraw-Hill Book Co., New York, 1953.
2. Birkhoff, Garrett, and Saunders MacLane, *A Survey of Modern Algebra*, Macmillan Co., New York, revised edition, 1953.
3. Dickson, L. E., *New First Course in the Theory of Equations*, John Wiley & Sons, New York, 1939.
4. Forsythe, G. E., Solving Linear Algebraic Equations Can Be Interesting, *American Mathematical Society, Bulletin*, vol. 59, p. 299, 1953.
5. Frazer, R. A., W. J. Duncan, and A. R. Collar, *Elementary Matrices and Some Applications to Dynamics and Differential Equations*, Cambridge University Press, Cambridge, 1947.
6. Halmos, Paul R., *Finite Dimensional Vector Spaces*, Princeton University Press, Princeton, 1942.
7. MacDuffee, C. C., *An Introduction to Abstract Algebra*, John Wiley & Sons, New York, 1940.
8. Milne, W. E., *Numerical Solution of Differential Equations*, John Wiley & Sons, New York, 1953.
9. Mitchell, B. E., Normal and Diagonalizable Matrices, *American Mathematical Monthly*, vol. 60, p. 94, 1953.
10. Riesz, F. and B. Sz-Nagy, *Functional Analysis*. Frederic Ungar Publishing Co., New York, 1955.
11. Schreier, O. and E. Sperner, *Introduction to Modern Algebra and Matrix Theory*, Chelsea Publishing Co., New York, 1951.
12. Sokolnikoff, I. S., *Tensor Analysis Theory and Applications*, John Wiley & Sons, New York, 1951.
13. Stoll, Robert R., *Linear Algebra and Matrix Theory*, McGraw-Hill Book Co., New York, 1952.
14. Thrall, R. M. and L. Tornheim, *Vector Spaces and Matrices*, John Wiley & Sons, New York 1957.
15. Turnbull, H. W. and A. C. Aitken, *An Introduction to the Theory of Canonical Matrices*, Blackie & Son Limited, Glasgow, 1945.
16. Whittaker, E. T., *A Treatise on the Analytical Dynamics of Particles and Rigid Bodies*, fourth edition, Dover Publications, New York, 1944.

Recommended Further Reading For The 2018 Edition

The importance of linear algebra in today's curricula has led to far too many textbooks in print at various levels to even begin to scratch the surface of here. The subject probably rivals calculus now for the sheer number of textbooks available. I will focus in this bibliography on books satisfying 2 criteria: a) Importance among standard texts by coverage and perceived quality and b) relative low cost. Of course, both are highly subjective. If your favorite isn't mentioned-well, tough toenails. Write your own damn bibliography.

The second edition of the great classic by Halmos (1) has been reissued recently in a beautiful low cost edition by Dover Books, much to the gratitude of the mathematical world. As described in the preface, it was effectively the first linear algebra textbook and it remains one of the best for advanced students. But as warned earlier, it is most definitely not an introduction. It is concise, somewhat dry and in typical Halmos fashion, many major results are presented as exercises. All that being said-it's beautifully written and crystal clear with a slant towards analysis many modern books lack. You won't find a better follow-up or supplementary text to Murdoch for the honors or advanced mathematics student.

Towards the other tail of the Gaussian bell curve of mathematical rigor in linear algebra texts is the equally famous textbook by Strang (2). I originally hesitated to add it simply because a new edition of this book is too damn expensive. In the end, I decided the book was so important and useful that to leave it out would simply be academic malpractice. Strang's lectures at MIT upon which this book is based have become legend since becoming available online at the OpenCourseWare site. This is a course built around the applications of linear algebra to real world problems. While this is true, it's somewhat misleading to say. It's hardly a plug-and-chug, mindless algorithm course: Strang analyzes each application and algorithm, as well as the theory behind it, thoroughly. But at the same time, it's not really a careful mathematics course the way we describe it. The deep theorems and proofs of linear algebra, while not ignored, are almost incidental here. The book intertwines the rigorous mathematics with the applications in such a way that the applications motivate the mathematics. Not every claim is proven, although most important results are. To the author, complete rigor is not as important in an introduction as the real world applications. For Strang, the complete abstract theory of linear algebra is really the function of an abstract algebra course. You can agree with this conclusion or not (I don't), but Strang's book has become a classic for good reason. You simply will not find a book that motivates the intuition behind the concepts of vector spaces and linear maps in a deeper and more comprehensive manner then Strang. You also won't find one with as many or as diverse applications of the subject as in this book: From classical geometry to differential equations to numerical approximation to game theory to linear programming and computer science, it's all here. It's important not to confuse this book with Strang's more recent, pared down texts on linear algebra such as *An Introduction To Linear Algebra*-this version is far more complete, although the others are more up-to-date. Strang himself has a wealth of additional material-video lectures, computer programming code, problem solutions, additional exercises and chapter extensions-at his website: http://math.mit.edu/~gs/ If you can afford both,I heartily recommend both the book and the site materials. If not, definitely check out the free materials at the website-especially Strang's wonderful recorded lectures.

Since one of our main missions here at BCS is to recommend low cost options, we turn now to low-cost linear algebra books that can be used as alternatives to Murdoch (although naturally, since all these books are so cheap, my intended recommendation is to buy several and use them all!)

As one would expect, Dover has republished a number of low cost linear algebra textbooks. Most of the early textbooks in linear algebra after Halmos' were written by Russian authors and Dover has republished English translations of many of them. All were translated by Richard Silverman. Most of them are quite good. They are distinctly "Russian" in style i.e. rigorous while avoiding abstraction and containing many applications to the physical sciences. They include Gel'fand (3), Smirnov(4) and 2 books by Shilov (5) and (6). (3) is written by one of the best Russian textbook authors, but it's not really a beginning text. I think many students might find parts of it tough going. Still, Gel'fand is a wonderful writer and if you can get a cheap copy, you might get quite a bit out of it. (4) isn't really a basic textbook on linear algebra, but it certainly contains a lot of useful information, including an introduction to one of the most important applications of linear algebra: group representation theory. But if I could only afford one, I'd go with (5), hands down. This by far is the best linear algebra book published by Dover besides (1). It's the best organized, clearest and most readable by far of all these books. It's also the most comprehensive. It would make wonderful collateral or subsequent reading to Murdoch.

Dover also publishes several American and UK linear algebra texts, such as Cullen (7), Damiano/Little (8) and Mirsky(9). All 3 are fairly standard linear algebra textbooks covering the standard topics. (9) is a bit more concise and difficult then (7) or (8). To be honest, I think Murdoch is better than all 3 of them. But you may disagree and the books are cheap, so by all means check them out and form your own opinion.

I'd be remiss if I didn't mention the Schaum's Outline on the subject (10). This is a very inexpensive, well written and comprehensive source that blends theory and applications well along with literally hundreds of detailed examples and solved problems. Indeed, I've known students over the years who have used it as their sole textbook for linear algebra and did quite well in the course. While I wouldn't go *that* far, you won't find a better study companion. It's certainly one of the very best Outlines. No student of linear algebra should enter any course without one-and keep it handy for review in subsequent courses.

Lastly, the recent trend of self-publication outlets like our own Createspace and others has resulted in the publication at least 2 original Open Source linear algebra texts. Both are available for free download and in inexpensive paperbacks: Hefferon(11) and Beezer (12). Both are very solid textbooks. The bulk of the material in either is pretty standard: Guassian and Guass-Jordan reduction of matrices, vector spaces, linear maps, determinants, and eigenvalues and eigenvectors. But both also have modern touches that make each unique in their own way. I'll close by describing and comparing both these books in some detail because they really are the first major original linear algebra textbooks available online. They demonstrate well what enormous potential this format has for new, original and inexpensive textbooks.

First Hefferon. While many universities have a separate "transition" course for mathematics majors that teach them the basic language and methods of proof in abstract mathematics after purely calculation courses, before this, certain keystone mathematics major courses served this purpose. The old "advanced calculus" class was the traditional course in which prior generations of mathematics majors made the adjustment to rigorous mathematics. More recently, the linear algebra course has filled that function. What makes Hefferon stand out is the fact the author has *intentionally* organized the text to specifically fulfill this dual purpose as both a first course in linear algebra and a first course in rigorous mathematical proof.

First of all, the book is considerably more precise then most books at this level are. There are basically 3 kinds of topics in the book: Definitions, theorems and applications. Topics that are usually treated algorithmically in books at this level are stated as theorems and proven. The discussion (and proof!) of the Guassian reduction process of linear systems in chapter 1 is a very good example. Examples are given as applications of theorems to specific cases. This is not to say this is a purely mathematical treatment like Halmos-a considerable number of applications of linear algebra to physical, social and computer science are discussed. Like Murdoch, applications to 2 and 3 dimensional classical geometry form a major part of Hefferon's book. But all applications are clearly shown to be derived consequences of proven results with a firm mathematical foundation. Secondly, there are a number of unusual applications that aren't usually present and all are carefully presented as consequences of proven results. To name just a few:

a) Kirchoff's circuit laws are treated as systems of n equations in n variables and solved accordingly for resistance, current or voltage,
b) the geometry of crystals is analyzed as a subspace of $\#^3$ with the minimum length of one side of a cubic crystal as the basis,
c) The so-called voting paradox is analyzed by taking linear combinations of counted voter preferences;
d) Laplace's and Chiò's methods for computing determinants of large n x n matrices,

e) Population stability analysis by solving for the eigenvalues of the recursive linear recursion relations.

While there's considerable overlap in content between Hefferon and Murdoch, there are many more modern applications in Hefferon that the student will not only find fascinating, but will demonstrate the enormous importance of this theory in many sciences beyond mathematics.

Beezer's book is considerably more unorthodox and creative then Hefferon in terms of content and in some regards, it's more advanced. First of all, the book is organized considerably differently. Hefferon is laid out in a very conventional numbered definition-theorem-proof-example format with the occasional breaking application. By contrast, Beezer's book is far less ordered in its organization, choosing instead to organize by content in a manner pioneered by Donald Knuth in his books. Every section, theorem, example, exercise and lemma is given an acronym that the student can use for either memorization, cross-referencing in the book or easy access in the online hyperlinked version. Some examples: The second chapter is simply called Chapter V because it discusses vectors, both geometric and algebraic. The theorem that proves the vector form of solutions to a linear system $\mathbf{AX} = \mathbf{B}$ is designated Theorem VFSLS. Personally, when reading it, I found this organizational system a bit confusing. That being said, I did find it more helpful when flipping around in the book cross referencing. Also, I'm sure when studying from the text, this acronym format might be quite useful for mnemonics. Secondly, the content is quite different in nature from either Murdoch or Hefferon. Classical geometry is almost completely excluded from the text and an un-apologetically algebraic approach is taken. Indeed, in many ways, Beezer's book is more mathematical and precise then either. Applications are limited almost entirely to computer science. The integration of the open source language Sage is quite interesting-like Hefferon, computer algorithms and code is used freely and frequently in the presentation. Unlike Hefferon, however, these sidebars into programming are limited to Sage implementations and theoretical computations rather than real world applications. All applications are purely mathematical. I'll allow the author to describe in his own words the structure of the text better than I can:

The first half of this text (through Chapter M) is a course in matrix algebra, though the foundation of some more advanced ideas is also being formed in these early sections (such as Theorem NMUS, which presages invertible linear transformations). Vectors are presented exclusively as column vectors (not transposes of row vectors), and linear combinations are presented very early. Spans, null spaces, column spaces and row spaces are also presented early, simply as sets, saving most of their vector space properties for later, so they are familiar objects before being scrutinized carefully……….. with a definition built on linear combinations of column vectors, it should seem more natural than the more frequent definition using dot products of rows with columns. And this delay emphasizes that linear algebra is built upon vector addition and scalar multiplication. Of course, matrix inverses must wait for matrix multiplication, but this does not prevent nonsingular matrices from occurring sooner. Vector space properties are hinted at when vector and matrix operations are first defined, but the notion of a vector space is saved for a more axiomatic treatment later (Chapter VS). Once bases and dimension have been explored in the context of vector spaces, linear transformations and their matrix representations follow. The predominant purpose of the book is the four sections of Chapter R, which introduces the student to representations of vectors and matrices, change-of-basis, and orthonormal diagonalization (the spectral theorem). This final chapter pulls together all the important ideas of the previous chapters. Our vector spaces use the complex numbers as the field of scalars. This avoids the fiction of complex eigenvalues being used to form scalar multiples of eigenvectors. The presence of the complex numbers in the earliest sections should not frighten students who need a review, since they will not be used heavily until much later, and Section CNO provides a quick review.... Sage is a powerful open source program for advanced mathematics. It is especially robust for linear algebra. We have included an abundance of material which will help the student (and instructor) learn how to use Sage for the study of linear algebra and how to understand linear algebra better with Sage. This material is tightly integrated with the web version of the book and will become even easier to use since the technology for interfaces to Sage continues to rapidly evolve. Sage is highly capable for mathematical research as well, and so should be a tool that students can use in subsequent courses and careers.

It's clear that Beezer's book is closer in spirit to Halmos' then many of the texts discussed here, although it is considerably gentler. The author cleverly works in quite sophisticated mathematical tools using only the language of linear algebra. For example, commutative diagrams are used sparingly but effectively. They come up most in the final chapter of the book, which discusses representation theory- certainly one of the most powerful and important applications of linear algebra in both pure and applied mathematics. It's also a notoriously difficult subject to present at this level, but Beezer does a solid job by using it as a capstone topic that utilizes virtually all the machinery developed in earlier chapters.

While I like both books, due to its straightforward organization and many applications, I think most beginning students would probably find Hefferon considerably more accessible and useful then Beezer. Like Halmos, I think Beezer's text would probably be better suited for either honors students or students in a second course.

References:

1) Halmos, Paul, *Finite Dimensional Vector Spaces*, 2nd edition, Dover Books, 2017
2) Strang, Gilbert, *Linear Algebra And Its Applications*, 4th edition, Brooks-Cole, 2006
3) Gel'fand, I. M., *Lectures On Linear Algebra*, Dover Books,
4) Smirnov, V.I., *Linear Algebra And Group Theory*, Dover Books,
5) Shilov, George, *Linear Algebra*, Dover Books,
6) Shilov, George, *Theory Of Linear Spaces*, Dover Books,
7) Cullen, Charles, *Matrices and Linear Transformations: Second Edition*, Dover Books,
8) Damiano, David B.; Little, John B.. *A Course in Linear Algebra*, Dover Books,
9) Mirsky, L. *An Introduction To Linear Algebra*, Dover Books,

10) Lipschutz, Seymour; Lipson, Marc, *Schaum's Outline Of Linear Algebra, Sixth Edition*, McGraw Hill, 2017

11) Hefferon, Jim, *Linear Algebra*, 3rd edition, Orthogonal Publishing L3C, 2017, http://joshua.smcvt.edu/linearalgebra/

12) Beezer, Rob, *A First Course In Linear Algebra*, Congruent Press, 2017, http://linear.pugetsound.edu

Answers to Problems

Exercise 1.2
4. $(7, -5, 9)$.   5. $(\frac{3}{5}, \frac{9}{5})$.   8. $11\sqrt{10}/6$.

Exercise 1.3
4. $18x + 8y - 11z = 0$.
5. $2x - y = 0$, $6x + z = 0$, or equivalent pair.
6. No.   7. $(39, -14, 30)$.

Exercise 1.4
1. Plane through the origin perpendicular to the vector $(1, -2, 1)$.
2. Space (a line) generated by $(1, 3, 5)$.
3. $(15, 1, -8)$.
7. (a) $(-1, -2, 1, 1)$.   (c) $(-1 - 13i, -5 - 3i, 11)$.

Exercise 1.5
1. (a) $S - 2T + U = 0$.   (b) $2S - U = 0$.
   (c) $5S - 2T + U = 0$.

Exercise 1.7
1. $(\frac{3}{5}, \frac{6}{5})$.   2. $(\sqrt{3}/2, 5\sqrt{3}/2)$.   3. $(-\frac{7}{5}, -\frac{18}{5}, \frac{17}{4})$.
5. $x'^2 + y'^2 + 2x'y' \cos(\alpha - \beta) = r^2$.
6. $4x'y' = a^2 + b^2$.

Exercise 1.8
1. $s + t \le n$.   2. $(5, -2, 3)$.

Exercise 1.9
4. $(2, -1, -1)$.
6. $(2/\sqrt{5}, 0, 1/\sqrt{5})$, $(-7/\sqrt{270}, -5/\sqrt{270}, 14/\sqrt{270})$, $(-1/\sqrt{54}, 7/\sqrt{54}, 2/\sqrt{54})$.

Exercise 1.10
2. (a) $5x'^2 - 6x'y' + 5y'^2 = 8$, $x'^2 + y'^2 = 1$.
   (b) $13x'^2 - 6\sqrt{3}x'y' + 7y'^2 = 16$, $x'^2 + y'^2 = 1$.

4. $x = x'' \cos\theta - y'' \sin\theta \cos\varphi + z'' \sin\theta \sin\varphi,$
$y = x'' \sin\theta + y'' \cos\theta \cos\varphi - z'' \cos\theta \sin\varphi,$
$z = y'' \sin\varphi + z'' \cos\varphi.$

Exercise 2.1

1. (a) 118. (b) −45.
4. Dependent if $x = 11$, independent for all other values of x.
5. $x = 1$.   7. $x = -\frac{13}{7}, y = -\frac{11}{7}$.

Exercise 2.2

1. (a) $x = z + \frac{10}{3}, y = -\frac{2}{3}$, z arbitrary.   (b) $x = y = 0$.
   (c) No solution.   (d) $z = t = x + y$, x and y arbitrary.
   (e) $x = 3, y = 0, z = -1$.
   (f) $x = \frac{1}{3}(5 - 3z), y = \frac{1}{2}(z - 3)$, z arbitrary.
7. $E_1 = \frac{1}{17}(7F_1 - 8F_2 + 6F_3), E_2 = \frac{1}{17}(-F_1 + 6F_2 - 13F_3),$
   $E_3 = \frac{1}{17}(3F_1 - F_2 + 5F_3).$
8. $X_1 = 14Y_1 - \frac{11}{3}Y_2 - \frac{5}{3}Y_3, X_2 = 17Y_1 - \frac{14}{3}Y_2 - \frac{5}{3}Y_3,$
   $X_3 = -11Y_1 + \frac{8}{3}Y_2 + \frac{4}{3}Y_3, Y_1 = \frac{43}{3}X_1 - \frac{11}{3}X_2 + \frac{8}{3}X_3,$
   $Y_2 = 44X_1 - 33X_2 + 5X_3, Y_3 = 23X_1 - 17X_2 + 3X_3.$

Exercise 3.1

1. $A^2 = \begin{pmatrix} 6 & -1 & -3 \\ 3 & 27 & 11 \\ -5 & 6 & 5 \end{pmatrix}$, $AB = \begin{pmatrix} 1 & 2 & 7 \\ 3 & -4 & 16 \\ -2 & -3 & -5 \end{pmatrix}$,

   $BA = \begin{pmatrix} -3 & 9 & 4 \\ -7 & -3 & 1 \\ 4 & 1 & -2 \end{pmatrix}$, $B^2 = \begin{pmatrix} 15 & 1 & -14 \\ 5 & 0 & -4 \\ 9 & 3 & 13 \end{pmatrix}$.

3. $x_1 = 7z_1 + 6z_2,$
   $x_2 = z_1 - 2z_2.$

5. $XY^t = (12), X^tY = \begin{pmatrix} 2 & 0 & -4 & 6 \\ -1 & 0 & 2 & -3 \\ 4 & 0 & -8 & 12 \\ 6 & 0 & -12 & 18 \end{pmatrix}$.

10. $PP^t = \begin{pmatrix} 21 & 0 \\ 0 & 30 \end{pmatrix}$, $P^tP = \begin{pmatrix} 13 & 4 & 5 & 12 \\ 4 & 5 & -6 & 8 \\ 5 & -6 & 17 & -4 \\ 12 & 8 & -4 & 16 \end{pmatrix}$.

Answers to Problems

Exercise 3.2

6. (a) $x/1 = y/8 = z/5$.   (b) $x/6 = y/(-3) = z/(-5) = t/(-7)$.
(c) $x/(-1) = y/2 = z/1 = t/1$.

7. $B = \begin{pmatrix} -7 & 0 & 0 \\ 5 & 0 & 0 \\ 13 & 0 & 0 \end{pmatrix}$, $C = \begin{pmatrix} 2 & 3 & -1 \\ 0 & 0 & 0 \\ 0 & 0 & 0 \end{pmatrix}$,

or $\quad B = C = \begin{pmatrix} 14 & 21 & -7 \\ -10 & -15 & 5 \\ -26 & -39 & 13 \end{pmatrix}$.

9. For example, $\begin{pmatrix} 1 & 2 & 1 \\ -1 & 0 & 1 \\ 0 & -1 & -1 \end{pmatrix}$ (answer not unique).

Exercise 3.4

1. (a) 3.   (b) 3.   (c) 2.
2. Dim. $\mathscr{S} = 2$, dim. $\mathscr{T} = 2$, dim. $\mathscr{S} + \mathscr{T} = 3$, dim. $\mathscr{S} \cap \mathscr{T} = 1$.
3. Rank = 3. Solution space generated by (0, 3, 2, 9).
4. Rank = 2.

Exercise 3.5

1. $P = \begin{pmatrix} 1 & 0 & 0 \\ 0 & 1 & 0 \\ -2 & -1 & 1 \end{pmatrix}$, $Q = \begin{pmatrix} 1 & 2 & -6 \\ 0 & 0 & 1 \\ 0 & -1 & 3 \end{pmatrix}$ (not uniquely determined).

Rank = 2.

Exercise 4.1

2. (a) Basis vectors $(-3, 0, 1), (2, 1, 0)$. Parametric equations $x = -3s + 2t$, $y = t, z = u$ (answers not unique).
(b) Basis vector $(1, -4, -3)$, $x = t, y = -4t, z = -3t$.
3. $11x - 67y + 31z + 48w = 0$.
6. $(\frac{18}{5}, -\frac{8}{15}, -\frac{28}{15})$.   7. $(\frac{9}{11}, -\frac{9}{11}, \frac{87}{11})$.
8. $(-\frac{142}{173}, \frac{133}{173}, \frac{124}{173}, \frac{105}{173})$ and $(\frac{420}{173}, \frac{188}{173}, -\frac{88}{173}, \frac{161}{173})$.

Exercise 4.2

1. (a) 10.   (b) 19.   (c) 10.
4. (a) $\sqrt{194}$.   (b) $3\sqrt{6}$.

Exercise 4.3

1. Vertices 16, edges 32, two-dim. faces 24, three-dim. faces 8.
2. $\dfrac{2^{n-r}n!}{r!(n-r)!}$.   3. (a) $\sqrt{4359}$. (b) 12.   4. $\sqrt{1049}$.   5. 27.

Linear Algebra for Undergraduates
Exercise 5.1

1. (a) $\begin{pmatrix} 2 & -1 \\ -3 & 2 \end{pmatrix}$, $\begin{aligned} x_1 &= 2y_1 + 3y_2, \\ x_2 &= y_1 + 2y_2. \end{aligned}$

   (b) $\begin{pmatrix} \frac{4}{7} & \frac{3}{7} \\ -\frac{2}{7} & \frac{12}{7} \end{pmatrix}$, $\begin{aligned} x_1 &= 2y_1 - \frac{y_2}{3}, \\ x_2 &= \frac{y_1}{2} - \frac{2y_2}{3}. \end{aligned}$

   (c) $\begin{pmatrix} \frac{4}{15} & \frac{4}{15} & -\frac{1}{15} \\ -\frac{7}{15} & \frac{8}{15} & -\frac{2}{15} \\ \frac{1}{4} & \frac{1}{4} & \frac{1}{4} \end{pmatrix}$, $\begin{aligned} x_1 &= 2y_1 + y_2 - 3y_3, \\ x_2 &= -y_1 + y_2, \\ x_3 &= y_2 + 4y_3. \end{aligned}$

   (d) $\begin{pmatrix} \frac{1}{8} & \frac{1}{8} & -\frac{1}{8} \\ \frac{4}{8} & -\frac{1}{8} & -\frac{5}{8} \\ -\frac{1}{8} & -\frac{1}{8} & \frac{7}{8} \end{pmatrix}$, $\begin{aligned} x_1 &= \tfrac{3}{8}y_1 + 4y_2 + y_3, \\ x_2 &= \tfrac{1}{2}y_1 - y_2, \\ x_3 &= \tfrac{1}{2}y_1 + y_3. \end{aligned}$

2. $\begin{pmatrix} \sqrt{3}/4 & -\sqrt{3}/4 \\ \sqrt{3}/2 & \sqrt{3}/2 \end{pmatrix}$ (OX' in first quadrant, OY' in third quadrant).

3. (1, 0, 1, 3).

Exercise 5.2

3. (a) $\begin{pmatrix} \frac{2}{3} & -\frac{2}{3} & \frac{1}{3} \\ \frac{2}{3} & \frac{1}{3} & -\frac{2}{3} \\ \frac{1}{3} & \frac{2}{3} & \frac{2}{3} \end{pmatrix}$. (b) $\begin{pmatrix} \frac{2}{7} & \frac{6}{7} & \frac{3}{7} \\ \frac{3}{7} & \frac{2}{7} & -\frac{6}{7} \\ \frac{6}{7} & -\frac{3}{7} & \frac{2}{7} \end{pmatrix}$. (c) $\begin{pmatrix} \frac{3}{5} & \frac{4}{5} & 0 \\ 0 & 0 & 1 \\ -\frac{4}{5} & \frac{3}{5} & 0 \end{pmatrix}$.

4. (a).

5. (a) $\begin{pmatrix} \frac{6}{7} & \frac{3}{7} & \frac{6}{7} \\ \frac{6}{7} & \frac{2}{7} & -\frac{3}{7} \\ \frac{3}{7} & -\frac{6}{7} & \frac{2}{7} \end{pmatrix}$. (b) $\begin{pmatrix} \frac{3}{5} & \frac{4}{5} & 0 \\ 0 & 0 & 1 \\ -\frac{4}{5} & \frac{3}{5} & 0 \end{pmatrix}$.

   (c) $\begin{pmatrix} -\frac{18}{35} & \frac{18}{35} & \frac{9}{7} \\ \frac{6}{7} & \frac{3}{7} & \frac{6}{7} \\ \frac{1}{35} & \frac{18}{35} & -\frac{6}{7} \end{pmatrix}$. (d) $\begin{pmatrix} \frac{18}{21} & \frac{13}{21} & -\frac{4}{21} \\ \frac{11}{21} & -\frac{16}{21} & -\frac{8}{21} \\ \frac{6}{21} & -\frac{4}{21} & \frac{19}{21} \end{pmatrix}$.

6. (c).

Exercise 6.1

1. (a) Magnification by a factor $|k|$ in direction of y-axis followed by a reflection in the x-axis if k is negative.
   (c) Magnification in every direction by a factor $|a|$ followed by a rotation through $180°$ (or reflection in both axes) if a is negative.
   (e) Rotation of the plane through $45°$.

Answers to Problems

2. (a) Kernel $2x + y = 0$, image space $y = 3x$. All points on the line $2x + y = k$ are mapped onto the point $(k, 3k)$.

3. $\begin{vmatrix} a-1 & b \\ c & d-1 \end{vmatrix} = 0.$

4. $(a-d)^2 + 4bc \geq 0.$ 5. $x' = 2x, y' = 3y.$
6. $x = 2(x' - y'), y = 3(x' + y').$
10. Determinant of its matrix $= \pm 1.$
12. $\pi ab.$ 13. $\pi abc.$

Exercise 6.2

1. τ_2^3 is $x' = -x, y' = -y, \tau_2^{12} = I.$

4. $P_{\sigma\tau} = \begin{pmatrix} \tfrac{2}{3} & \tfrac{13}{3} \\ -\tfrac{1}{3} & \tfrac{2}{3} \end{pmatrix}$, $P_{\tau\sigma} = \begin{pmatrix} \tfrac{2}{3} & \tfrac{1}{3} \\ -\tfrac{13}{3} & \tfrac{2}{3} \end{pmatrix}.$

5. $A_\tau = \begin{pmatrix} 2 & 5 \\ -1 & 6 \end{pmatrix}$, $A_{\tau^{-1}} = \begin{pmatrix} \tfrac{6}{17} & -\tfrac{5}{17} \\ \tfrac{1}{17} & \tfrac{2}{17} \end{pmatrix},$

relative to both bases.

6. (a) $\begin{pmatrix} \tfrac{8}{7} & \tfrac{10}{7} \\ -\tfrac{1}{7} & -\tfrac{5}{7} \end{pmatrix}.$ (b) Same as (a).   (c) $\begin{pmatrix} \tfrac{11}{7} & 1 \\ -\tfrac{5}{7} & -1 \end{pmatrix}.$

Exercise 6.3

1. (a) $\begin{pmatrix} 7 & -6 \\ -5 & 5 \end{pmatrix}.$ (b) $\begin{pmatrix} 0 & 1 \\ -5 & 12 \end{pmatrix}.$ (c) $\begin{pmatrix} 1 & \tfrac{1}{2} \\ 12 & 11 \end{pmatrix}.$

2. $\begin{pmatrix} -3 & 0 \\ 0 & 2 \end{pmatrix}$; reflection in the line $x + y = 0$ followed by a magnification by factor 3 along the line $x - y = 0$ and by a factor 2 along the line $x + y = 0.$

5. $\begin{pmatrix} 8 & 6 \\ -2 & 0 \end{pmatrix}.$ 6. $\begin{pmatrix} 3 & -\tfrac{2}{3} & -\tfrac{2}{3} \\ -\tfrac{2}{3} & \tfrac{10}{3} & 0 \\ -\tfrac{2}{3} & 0 & \tfrac{8}{3} \end{pmatrix}.$

8. (a) Factor 2 in direction of $(1, -2)$ and factor 5 in direction of $(1, 1).$

Exercise 6.4

3. (a) $\begin{pmatrix} (8 + 5\sqrt{2})/18 & (8 - \sqrt{2})/18 & (2 - 4\sqrt{2})/9 \\ (8 - 7\sqrt{2})/18 & (8 + 5\sqrt{2})/18 & (2 + 2\sqrt{2})/9 \\ (2 + 2\sqrt{2})/9 & (2 - 4\sqrt{2})/9 & (1 + 4\sqrt{2})/9 \end{pmatrix}.$

www.ingramcontent.com/pod-product-compliance
Lightning Source LLC
Chambersburg PA
CBHW052310220526
45472CB00001B/55